Tricia Carmichael, Hyun-Joong Chung (Eds.)
Stretchable Electronics

Also of Interest

Nanoelectronics
From Device Physics and Fabrication Technology to Advanced Transistor Concepts
2nd Edition
Joachim Knoch, 2024
ISBN 978-3-11-105424-7, e-ISBN (PDF) 978-3-11-105442-1,
e-ISBN (EPUB) 978-3-11-105501-5

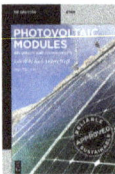

Photovoltaic Modules
Reliability and Sustainability
2nd Edition
Karl-Anders Weiß (Ed.), 2021
ISBN 978-3-11-068554-1, e-ISBN (PDF) 978-3-11-068555-8,
e-ISBN (EPUB) 978-3-11-068572-5

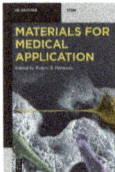

Materials for Medical Application
Robert B. Heimann (Ed.), 2020
ISBN 978-3-11-061919-5, e-ISBN (PDF) 978-3-11-061924-9,
e-ISBN (EPUB) 978-3-11-061931-7

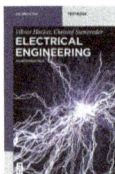

Electrical Engineering
Fundamentals
Viktor Hacker, Christof Sumereder, 2020
ISBN 978-3-11-052102-3, e-ISBN (PDF) 978-3-11-052111-5,
e-ISBN (EPUB) 978-3-11-052113-9

Photovoltaic Module Technology
2nd Edition
Harry Wirth, 2020
ISBN 978-3-11-067697-6, e-ISBN (PDF) 978-3-11-067701-0,
e-ISBN (EPUB) 978-3-11-067710-2

Stretchable Electronics

The Next Generation of Emerging Applications

Edited by
Tricia Carmichael and Hyun-Joong Chung

DE GRUYTER

Editors
Professor Tricia Carmichael
University of Windsor
Department of Chemistry and Biochemistry
401 Sunset Avenue
Windsor, Ontario N9B 3P4
Canada
tbcarmic@uwindsor.ca

Professor Hyun-Joong Chung
University of Alberta
Department of Chemical and Material engineering
9211 116 Street NW
Edmonton, Alberta T6G 1H9
Canada
chung3@ualberta.ca

ISBN 978-3-11-075718-7
e-ISBN (PDF) 978-3-11-075728-6
e-ISBN (EPUB) 978-3-11-075731-6

Library of Congress Control Number: 2024943393

Bibliographic information published by the Deutsche Nationalbibliothek
The Deutsche Nationalbibliothek lists this publication in the Deutsche Nationalbibliografie;
detailed bibliographic data are available on the Internet at http://dnb.dnb.de.

www.degruyter.com

Contents

Gloria M. D'Amaral, Hannah R. Jessop, and Tricia Breen Carmichael
Beyond Moore's law to more than Moore in stretchable electronics

Abstract: Over the past two decades, the field of stretchable electronics has advanced from initial ideas exploring the mechanics of simple metal-on-elastomer structures to complex, soft, and wearable technologies with applications in human health, robotics, energy storage, and more. This chapter acquaints the reader with this exciting and fast-paced field, focusing on the innovative design elements developed over the years to endow functional materials with stretchability. These design elements are divided into two philosophies: Structures that stretch uses intentional architectural designs—from simple wavy structures to highly complex hierarchical arrangements—that convert stretching strains into non-destructive bending strains. Materials that stretch takes a materials-based approach to produce intrinsically stretchable functional materials through compositing or molecular engineering. This chapter provides the reader with the fundamental information necessary to understand the latest exciting developments in the field.

Wearable electronic devices have existed since the turn of the last century, beginning with the introduction of wristwatches in 1904. Over the next ~100 years, wristwatches evolved into sophisticated, multifunctional smartwatches capable of not only keeping time, but also monitoring a user's health and fitness, tracking sleep, measuring environmental sound levels, and more (Figure 1a). This incredible progress has been the result of the phenomenal advances in semiconductor technologies. Moore's Law has been the primary driver of this progress over the past 50 years, continually improving the performance-to-cost ratios of products and inducing exponential growth of the semiconductor market. What the International Technology Roadmap for Semiconductors (ITRS) termed "More Moore" continues today. However, a new vision for wearables looks beyond Moore's Law. "More than Moore" (MtM) was added to the ITRS in 2005 [1]. MtM is the functional diversification of semiconductor-based devices that broadens the roadmap to include a forecast for devices designed to interact with people and the environment (Figure 1b). MtM recognizes the modern importance of device design driven by application needs, rather than the technology requirements of More Moore. Wearable devices constitute a major part of MtM, with application needs that go beyond device functionality to include the important question of wearability. Put simply, while More Moore focuses on the continual improvement of device performance, MtM asks an entirely different question: *What devices do we really want to wear?*

Gloria M. D'Amaral, Hannah R. Jessop, Tricia Breen Carmichael, Department of Chemistry and Biochemistry, University of Windsor, 401 Sunset Ave., Windsor, Ontario N9B 3P4, Canada, e-mails: damaral@uwindsor.ca, jessoph@uwindsor.ca, tbcarmic@uwindsor.ca

https://doi.org/10.1515/9783110757286-001

a

b

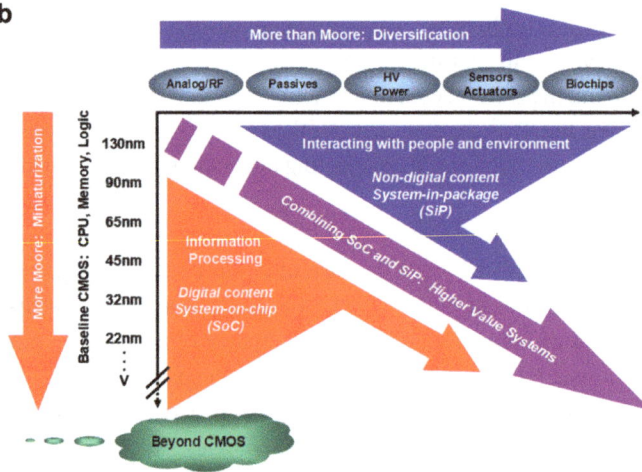

Figure 1: (a) Timeline depicting the development of wearable displays. Retrieved and used with permission from https://commons.wikimedia.org/w/index.php?title=File:Wrist_Watch_WWI.jpg&oldid=681778140; Fancycrave1 2015 Apple Smart Watch https://pixabay.com/en/smart-watch-apple-technology-style-821563/ (CC BY 1.0); reference [2]. (b) Schematic illustration "More Moore" and "More than Moore" from the International Technology Roadmap for Semiconductors (ITRS) 2005. Reproduced with permission from reference [1].

The vision for wearability that emerged in the early 2000s focused on the integration of electronics with soft and stretchable substrates to match the mechanical properties of the human body. Soft devices can be seamlessly integrated with the body, either on the skin or implanted within. The progression of wristwatches in Figure 1a illustrates this research direction with an example of a soft, lightweight wearable display that conforms to human skin. However, integrating functional electronic materials with soft, stretchable substrates is a big challenge. Materials used in electronics, such as metals and silicon, are typically not stretchable. For example, a gold film deposited on the elastomer polydimethylsiloxane (PDMS) by electron beam evaporation rapidly develops long chan-

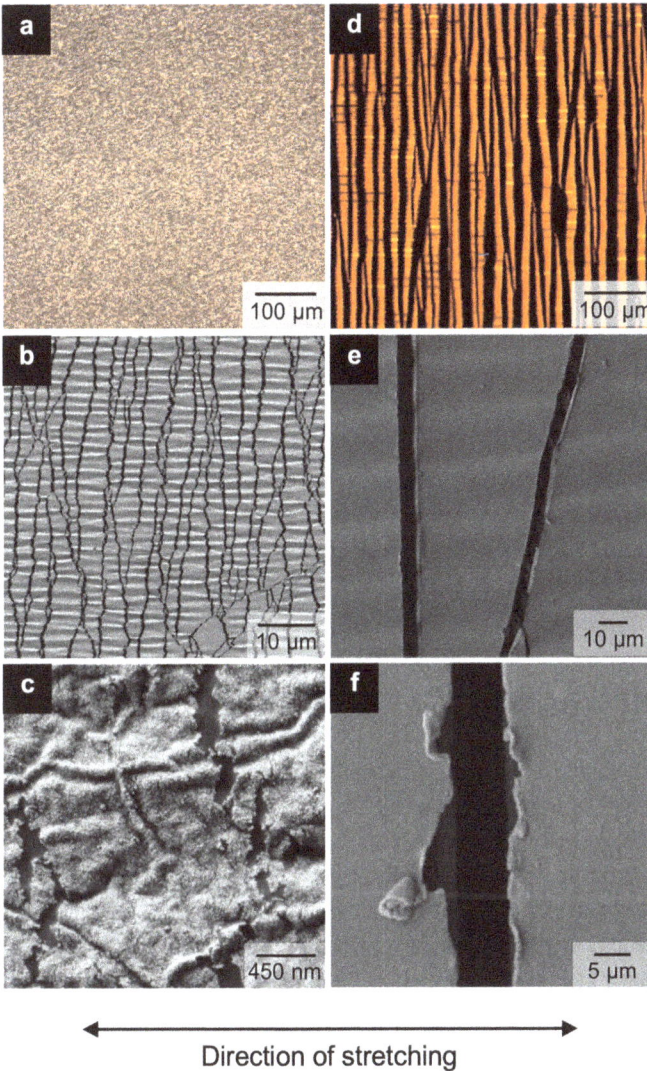

Direction of stretching

Figure 2: Engineering the cracking patterns of gold films on PDMS. Optical (top) and SEM (middle, bottom) images of gold films with (a–c) a heterogeneous crystalline surface texture and (d–f) a dominant [111] texture at 30 % elongation (a, d) and 30 % elongation (b, c, e, f). Reproduced with permission from reference [10].

nel cracks with stretching to relieve strain, breaking the conductive pathway [2]. Even functional organic polymers do not guarantee stretchability, instead exhibiting a wide variety of mechanical properties that are dramatically influenced by subtle differences in chemical structure [3]. Engineering the cracking patterns in film-on-elastomer systems is one approach to handling the modulus mismatch between functional materials and elastomers [4]. This approach aims to avoid the formation of destructive channel cracks by instead inducing a pattern of fine cracks in the film to distribute strain relief, preserve conductive pathways, and thus extend the functional range of the film. Methods to engineer crack formation include increasing the roughness of the elastomeric substrate to provide numerous sites for strain localization and crack initiation [5–8] and manipulating the crystalline orientation of the overlying metal film to create misaligned grain boundaries that impede crack propagation (Figure 2) [9]. Crack engineering methods typically lead to an increase in the resistance of the film with stretching, which can be exploited to fabricate soft and wearable resistive strain sensors [10]. However, finding ways to solve the modulus mismatch between functional materials and elastomers without damage to the film has driven much of the progress in stretchable electronics, leading to structures and materials that have been used as design elements in various approaches to fabricate advanced wearable devices. In this chapter, we discuss these design elements that continue to be used to advance stretchable electronics. These design elements come in two categories: 1) structures that stretch, in which materials that are not intrinsically stretchable are configured into shapes that convert tensile strain into less-destructive bending strain; and 2) materials that stretch, in which functional materials and composites are developed that are intrinsically stretchable.

1 Structures that stretch

Solid materials become flexible if they are configured into sufficiently thin forms, typically between ~100 nm and ~1 µm, due to the linear decrease in bending strain with thickness [11]. This mechanical principle underpins a variety of stretchable structures that have played an integral role in launching stretchable electronics. Architectural designs that incorporate wavy structures—both in plane and out of plane—convert tensile strains into less destructive bending strains of individual architectural features [12]. Wavy designs have played a key role in advancing stretchable electronics, enabling the field to move beyond structures that stretch to complex stretchable circuits and systems. One of the most influential outcomes of wavy designs is the island–bridge architecture, a design strategy in which conductive wavy traces (bridges) interconnect rigid functional components (islands) [13]. This design provides elastic circuits that isolate the functional components from strain, with the bridge structures accommodating nearly all the deformation. Island–bridge designs accommodate stretching without compromising functionality by enabling the use of conventional microfabrication techniques on rigid substrates to fabricate high-performance device "islands".

1.1 In-plane patterned wavy structures

In 2004, Gray et al. described the fabrication of thin metal films lithographically patterned into serpentine shapes on the surface of a PDMS substrate [14]. The serpentine

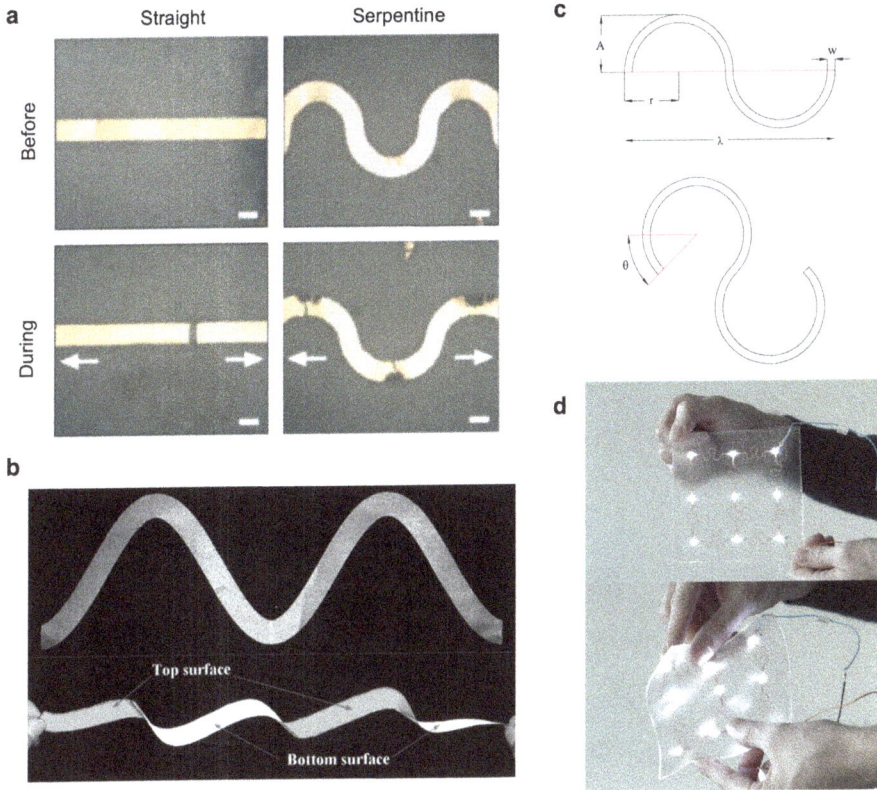

Figure 3: Serpentine interconnects. (a) Optical micrographs comparing straight and serpentine gold wires on PDMS and before and during stretching. Arrows indicate the direction of stretching. Reproduced with permission from reference [15]. (b) Stretching a serpentine-shaped piece of paper to demonstrate the out-of-plane twisting with elongation. Reproduced with permission from reference [16]. (c) Geometric parameters of serpentine (top) and horseshoe (bottom) wires. Reproduced with permission from reference [5]. (d) Stretchable a 3 × 3 array of LEDs with serpentine interconnects. Reproduced with permission from reference [25].

wires in this study remained conductive up to 27 % elongation, while straight wires of the same width on PDMS failed electrically at 2 % strain due to cracking (Figure 3a). Stretching the PDMS substrate causes the serpentine structures to bend and twist out of plane at the peaks and troughs to accommodate the strain, like a two-dimensional spring. Pulling a piece of paper cut into a serpentine shape illustrates the out-of-plane deformation that must be accommodated by the compliant substrate (Figure 3b) [15]. Adhesion of the serpentine material to the compliant substrate is key: Delamination under strain produces locally free-standing areas that readily fracture. Serpentines that are well-adhered to highly compliant substrates are thus ideal and can maintain functionality until the serpentine becomes straightened, after which point fracture will occur [15, 16]. Numerous subsequent studies have explored the effect of geometric parameters

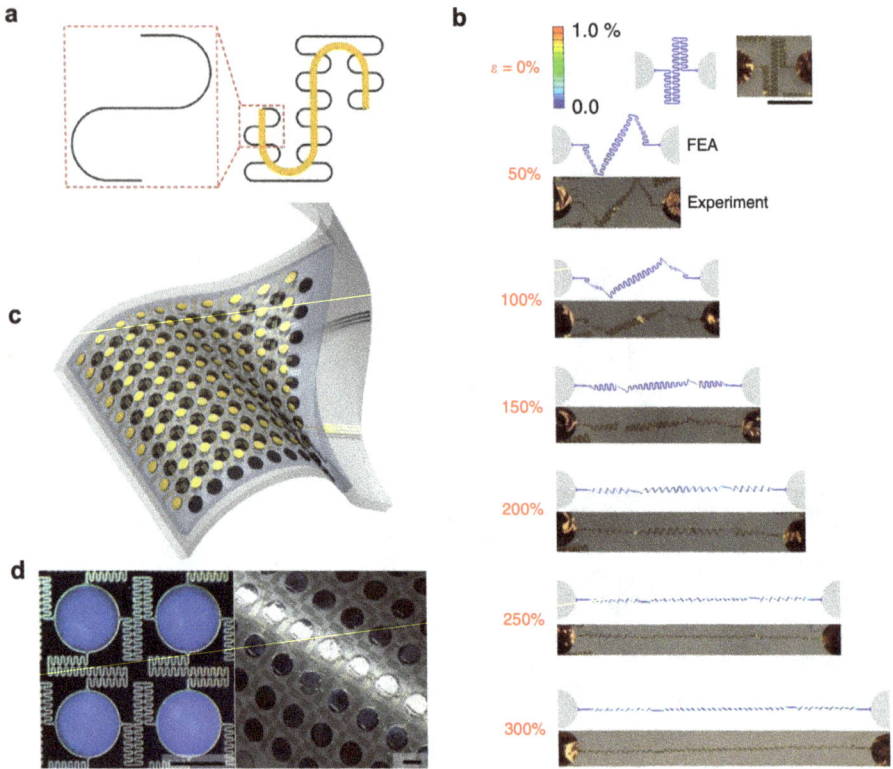

Figure 4: Self-similar serpentine interconnects. (a) Illustration of self-similar serpentine geometries. (b) Optical images and finite element analysis (FEA) of deformation modes from 0–300 % strain (scale bar = 2 mm). (c) Illustration of stretchable lithium-ion battery array with self-similar serpentine interconnects. (d) Optical images of the array on a silicon wafer (left) and after transference to an elastomeric substrate. Reproduced with permission from reference [27].

of serpentine shapes, such as amplitude (A), wavelength (λ), radius (r), width (w), and the angle θ (Figure 3c) to maximize stretchability, while preserving stable electrical conductivity [17–19]. Incorporating redundancy into the design by fabricating multiple parallel serpentine wires also improves performance [14, 16, 20, 21]. Optimized structures can withstand strains as high as hundreds of percent elongation [22, 23].

The island–bridge architecture with simple serpentine interconnects is illustrated in the stretchable matrix of light tiles in Figure 3d [24]. This image highlights the low areal fill factor (i. e., the ratio between the actual area of the substrate and the area occupied by the devices) that are a consequence of using serpentine interconnects in the island–bridge design. Serpentine interconnects can also exhibit sharp strain concentrations at specific points with deformation, which can limit performance in stretchable electronics applications [25]. Furthermore, simple serpentine structures accommodate stretching mainly along the long axis of the wave, with limited stretchability normal to

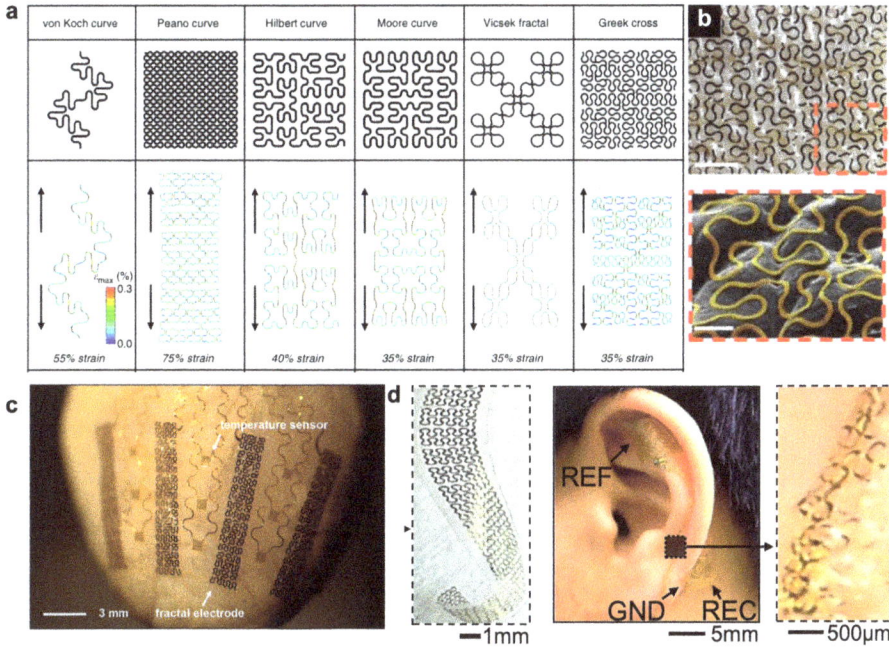

Figure 5: Fractal interconnects. (a) Examples of fractal layouts (top) with corresponding FEA images of each structure under tensile strain. (b) Optical (top) and SEM images (bottom) of Peano-based fractal wires in conformal contact with skin and a skin-replica, respectively (scale bars are 2 mm and 500 mm, respectively). (c) Stretchable low-impedance fractal electrodes for electrical stimulation in conformal contact with a Langendorff-perfused rabbit heart. (d) Fractal interconnects used in stretchable EEG measurement system, with photographs of the EEG device laminated on the auricle of a human ear. Reproduced with permission from (a, b) reference [30], (c) reference [31], and (d) reference [32].

the long axis. For truly wearable electronics, structures that can accommodate omnidirectional stretching are needed to better conform to complex motions of the human body. This need has inspired more sophisticated serpentine designs and layouts, such as self-similar serpentines, fractal-inspired designs, and spiral designs capable of these deformations [12]. Adding greater complexity to the design leads to structures that not only accommodate omnidirectional stretching but also provide greater stretchability, as well as higher areal fill factors, than conventional serpentine structures.

Self-similar serpentines are space-filling interconnect designs comprising two-level wave-shape structures, a "serpentine of serpentines", which exhibit two levels of unraveling with strain [26]. Figure 4a illustrates this concept, which takes the serpentine unit cell illustrated schematically in the red box, reduces the scale, and then connects multiple copies of it in the layout of the original unit-cell geometry [27]. The deformation mechanism involves initially unraveling the 'long' wavelength serpentines, while the 'short' wavelength serpentines remain relatively undeformed. After the 'long' wavelength serpentines have fully extended, the 'short' wavelength structures unravel with

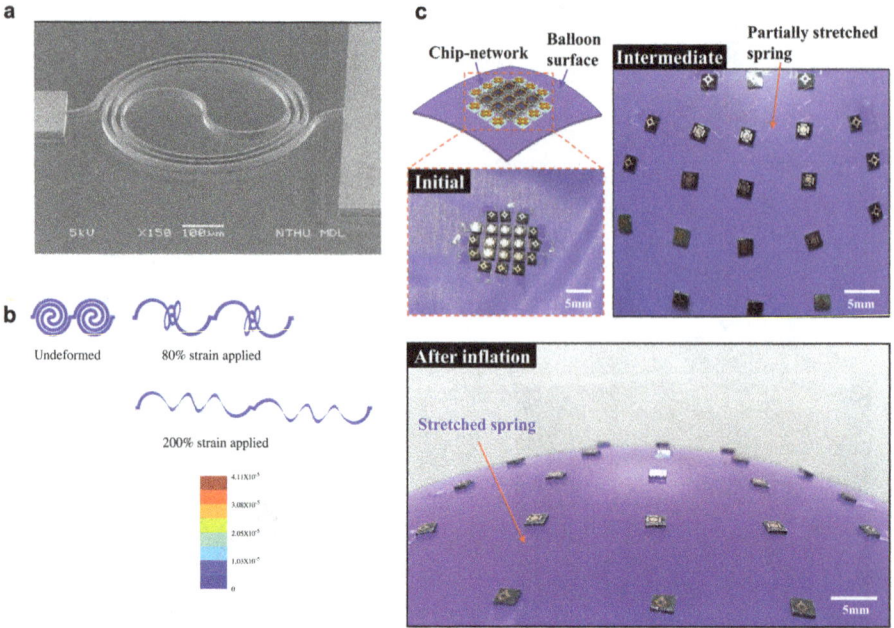

Figure 6: 2D spiral interconnects. (a) SEM image of copper spiral interconnect fabricated on silicon. (b) FEA images of 2D spiral at 0 % strain and stretched to 80 % and 200 % strain. The color map indicates the low plastic deformation. (c) 2D array of rigid sensors with copper 2D spiral interconnects on the surface of a balloon in the deflated, intermediate, and fully expanded state where the spring is stretched to its full length. Reproduced with permission from (a, c) reference [33], (b) reference [34].

additional stretching (Figure 4b) [28]. This design strategy improves the areal fill factor and stretchability by making use of the empty space between adjacent islands to increase the total length of the interconnects. For example, an array of lithium-ion batteries interconnected by self-similar serpentines exhibit ordered unraveling to >300 % biaxial stretchability with areal fill factors as high as 50 % (Figure 4c, d) [27].

Fractal-based structures take these self-similar designs to the next level, engineering patterns that accommodate biaxial strain, radial strain, or even strain in a particular direction [29]. Figure 5a shows examples of topologies, such as lines (Koch, Peano, Hilbert), loops (Moore, Vicsek), and branch-like meshes (Greek cross). The deformation of these structures has been validated experimentally and through the finite element method (FEM). Fractal-based structures, by virtue of their self-similar geometric patterns repeating at different scales, distribute strain more uniformly compared to simple interconnect designs like serpentines. Fractal-inspired structures can be designed to conform to the fine features of human skin and other complex biological surfaces (Figure 5b–d) [30, 31].

Two-dimensional spiral-shaped structures are distinct from the serpentines and fractals in their mode of deformation, which involves an unwinding of the spiral through out-of-plane twisting and bending deformations (Figure 6a, b) [32]. This unwinding process allows 2D spirals to accommodate strain energy effectively, supporting omnidirectional stretching to high stretchabilities of >200 % [32, 33]. The great advantage of 2D spiral structures, however, is that they can create strain-free connections between rigid device islands [12]. This feature makes island–bridge stretchable electronics resilient to significant mechanical stresses and deformations. Furthermore, the compact design can provide high areal densities (Figure 6c) [34].

1.2 Out-of-plane wavy structures

Wavy structures that protrude out of the plane of the elastomeric substrate use the same underlying mechanism of bending of thin structures for stretchability. Initial demonstrations of this strategy in the early 2000s used a straightforward prestrain strategy to achieve out-of-plane deformation [35, 36]. This strategy involves mechanically prestretching a PDMS substrate uniaxially, biaxially, or radially, and then depositing a functional material such as a metal film on the surface. Releasing the prestrain leads to a buckling instability due to the stiffness of the overlying metal film compared to that of the elastomeric substrate, which induces the formation of an out-of-plane wavy structure (Figure 7a) [37, 38]. The minimum prestrain required to generate buckle formation (ε_c) depends on the elastic moduli (E) and the Poisson ratio (v) of the overlying film (f) and elastomeric substrate (s) according to Equation (1):

$$\varepsilon_c = \frac{1}{4}\left(\frac{3E_s/(1-v_s^2)}{E_f/(1-v_f^2)}\right)^{2/3}. \tag{1}$$

After releasing the prestrain, stretching the wavy structure releases the built-in compression by unbending and flattening the buckles. Although it may seem like a good

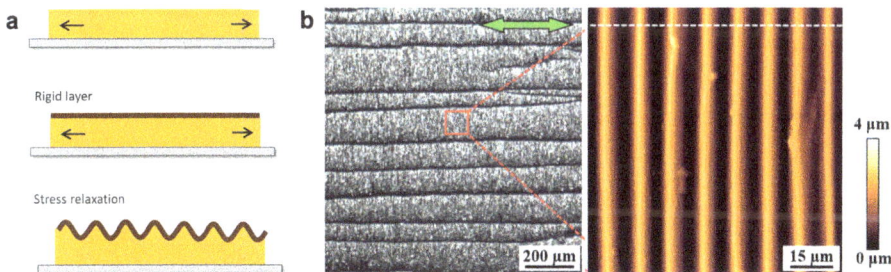

Figure 7: Prestrained metal films on PDMS. (a) Schematic of the prestrain method to create buckled thin films on elastomers. (b) SEM (left) and AFM (right) images of buckled silver films on PDMS formed upon releasing the prestrain. The bidirectional arrow represents the prestrain loading direction. The SEM image shows the cracks that form parallel to the loading direction due to the Poisson effect. Reproduced with permission from (a) reference [38], (b) reference [40].

idea to maximize the magnitude of the prestrain to also maximize the working range of the overlying film, it is important to consider the consequence of the Poisson effect. For the simple case of uniaxial strain, the prestrained state bears tensile strain parallel to the prestrain direction and compressive strain perpendicularly due to the Poisson effect. Releasing the prestrain applies compressive stress to the overlying film in the prestrain direction (leading to buckles), but also tensile strain perpendicularly due to the Poisson effect, which can cause cracks to form in the metal film parallel to the prestrain direction (Figure 7b) [39]. In practice, prestrained metal/PDMS structures can remain conductive to strains as high as ~100 %, equivalent to the initial prestrain value.

Even silicon, a brittle material, can become stretchable with the prestrain approach. Silicon nanomembranes with thicknesses of 10–100 nm fabricated from silicon-on-insulator wafers can be adhered to a PDMS support. Heating used in the fabrication process produces a controlled degree of isotropic thermal expansion of the PDMS substrate. Subsequent cooling releases the thermally induced prestrain, leading to the formation of buckled silicon nanomembranes with herringbone patterns (Figure 8a) [40]. These structures exhibit full 2D stretchability that transform the herringbone structures depending on the direction of uniaxial tensile strain (Figure 8b). As the herringbone structures "unfold" with strain, the Poisson effect induces compressive strain

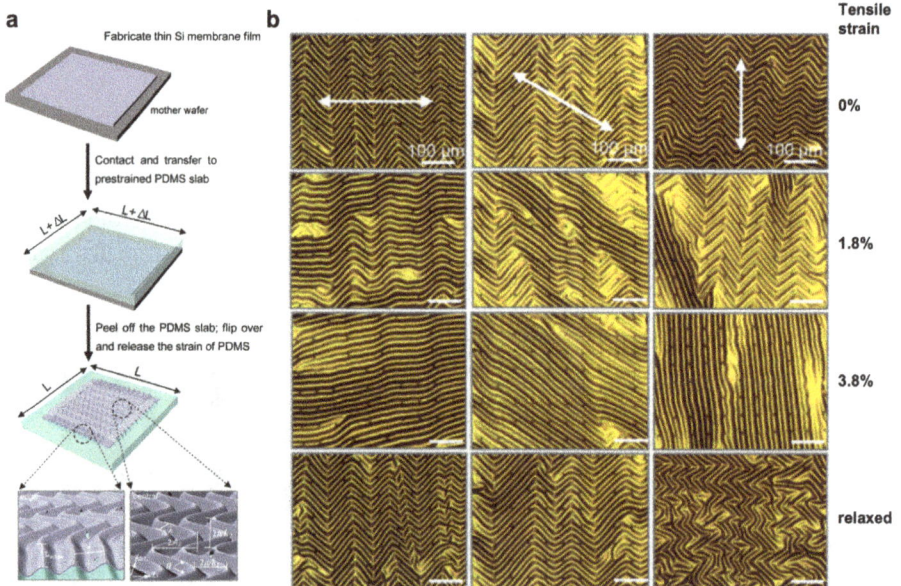

Figure 8: Prestrained silicon films on PDMS. (a) Schematic of the process for fabricating a wavy silicon nanomembrane on an elastomeric substrate. (b) Optical micrographs of wavy silicon nanomembranes under various uniaxial strains applied at different orientations indicated by the bidirectional arrows. The images were collected in the relaxed state before stretching (top), at uniaxial strains of 1.8 % and 3.8 % (middle), and in the relaxed state after stretching (bottom). Reproduced with permission from reference [41].

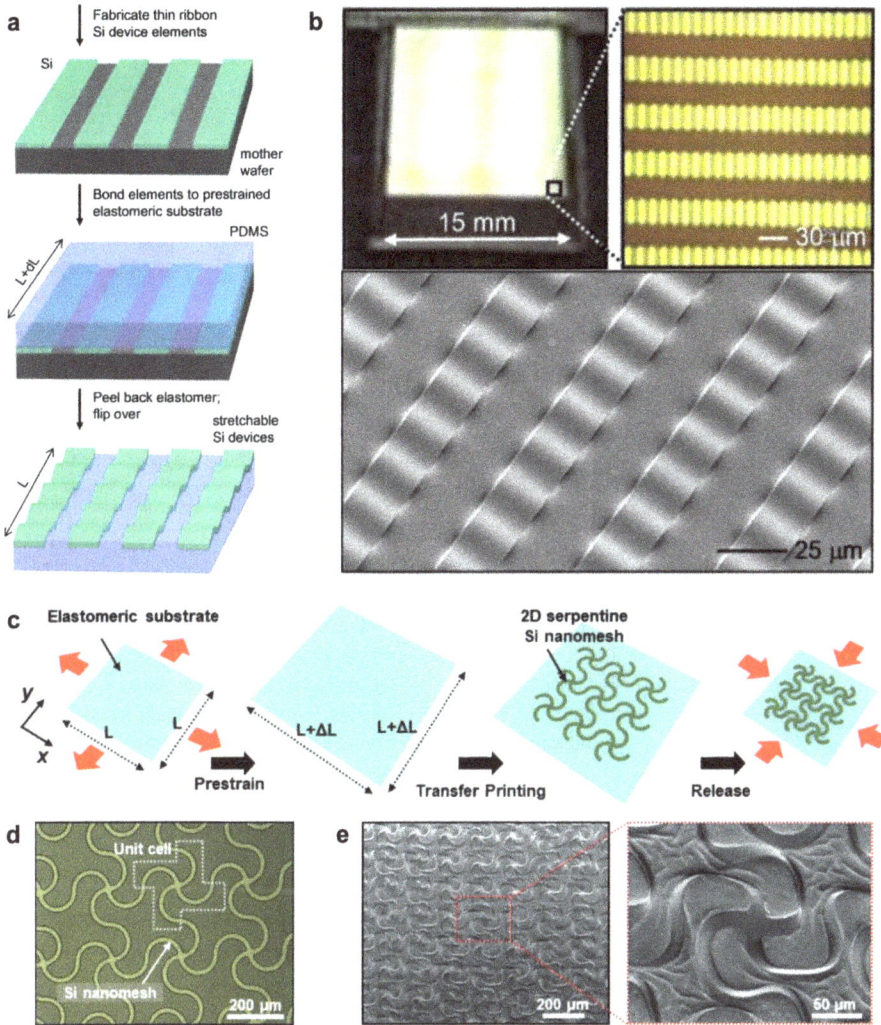

Figure 9: Prestrained silicon nanoribbons on PDMS. (a) Schematic illustration of the fabrication process for stretchable single-crystal silicon nanoribbons on elastomeric substrates. The nanoribbons are first fabricated on a silicon-on-insulator (SOI) wafer and then bonded to a prestrained elastomeric substrate. (b) Optical images (top) and SEM images (bottom) of a large-scale aligned array of wavy, single-crystal Si ribbons on PDMS. (c) Schematic illustration of the fabrication process for biaxially stretchable serpentine silicon nanomesh. (d) Optical microscope image of fabricated serpentine silicon nanomesh on an SOI wafer. (e) SEM image of the nanomesh on an elastomeric substrate after releasing the prestrain. Reproduced with permission from (a, b) reference [42] and (c–d) reference [44].

in the orthogonal direction. Unlike the uniaxial prestrained structures just described, compressive strain in herringbone structures can be accommodated by compression of the wavy structures in this direction.

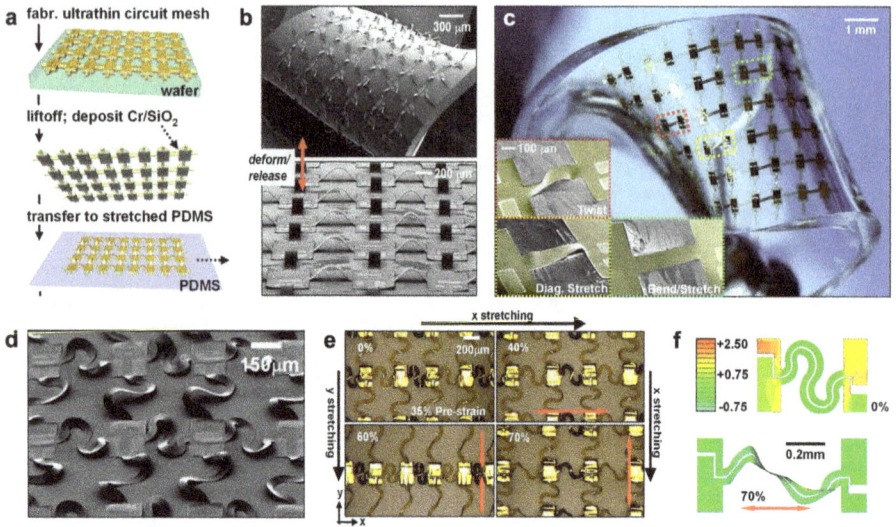

Figure 10: Noncoplanar stretchable structures with patterned adhesion. (a) Schematic illustration of the fabrication process of a circuit mesh on a prestrained PDMS substrate with selective adhesion at the nodes. (b) SEM images of an array of CMOS inverters showing the out-of-plane structure in an undeformed state (lower) and in a twisted configuration (upper). (d) SEM image of an array of stretchable CMOS inverters with noncoplanar serpentine-shaped interconnects. (e) Optical images of stretching the array in the *x* and *y* directions. (f) Finite element modeling (FEM) simulation before (left, after releasing 35 % prestrain) and after at 70 % tensile strain. Reproduced with permission from reference [45], Copyright (2008) National Academy of Sciences, U. S. A.

Patterning silicon nanomembranes into ribbon-shaped structures with widths in the μm range incorporates tolerance of the Poisson effect into the design (Figure 9a) [41, 42]. Silicon nanoribbons bonded to a uniaxially prestrained PDMS substrate spontaneously form well-controlled, highly periodic sinusoidal structures upon relaxation of the prestrain (Figure 9b) [41]. This stretchable form of silicon remains intact and functional beyond the intrinsic fracture limit of silicon (~1 %). For example, these structures can accommodate strains as high as 5.5 % when a prestrain value of 3.5 % is used [41]. At the same time, the PDMS regions between the nanoribbons absorb the perpendicular strains caused by the Poisson effect, contributing to the overall mechanical performance.

Taking this idea a step further, a combined strategy of prestrain and serpentine shapes provides silicon structures with remarkable stretchability and higher areal fill factors. Silicon nanoribbons configured into a serpentine-shaped nanomesh on a biaxially prestrained PDMS substrate forms a compressed structure by bending either inward or outward when the prestrain is released (Figure 9c–e) [43]. With subsequent tensile strain, the serpentine nanomesh initially flattens, and then the serpentine structure allows it to accommodate further applied strain without causing cracks or fractures.

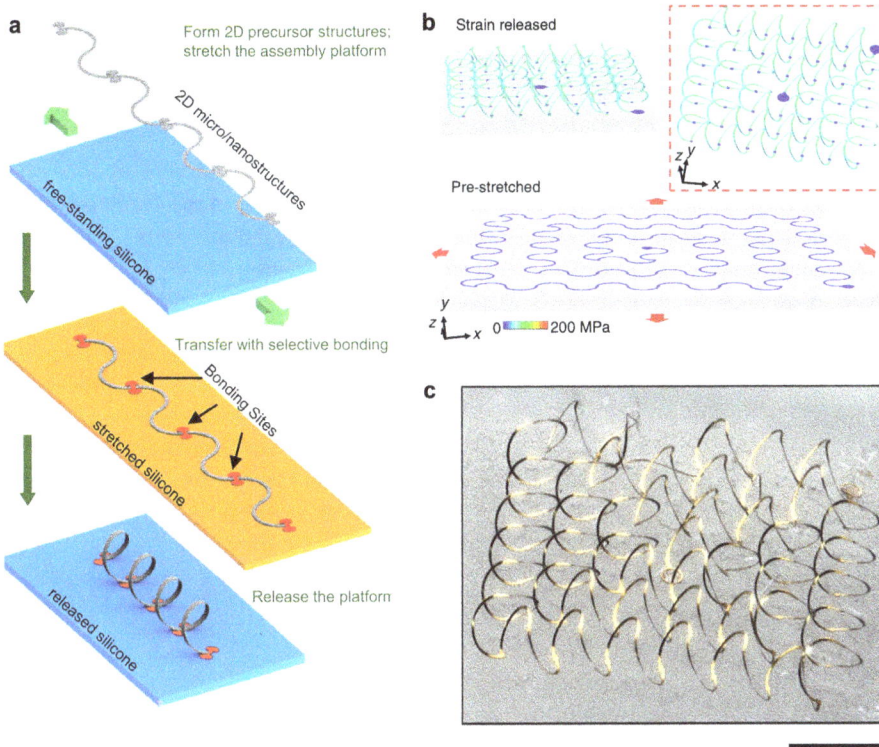

Figure 11: 3D helical structures. (a) Schematic illustration of the assembly process showing the selective bonding sites in red and formation of 3D spring-like structure upon releasing the prestrain. (b) FEA depiction of the formation of a 3D network of helical interconnects. The color represents the magnitude of Mises stress in the metal layer. (c) Optical image of the 3D network of metal helices bonded to a silicone substrate (scale bar, 1 mm). Reproduced with permission from (a) reference [51], (b, c) reference [48].

Serpentine-shaped silicon nanoribbons on a PDMS substrate with a biaxial prestrain of 50 % can be stretched by an impressive 75 % biaxially, illustrating the power of the combined effect. The combined strategy also results in an areal fill factor that is 125 % higher compared to that of a serpentine layout without prestrain.

These approaches for achieving stretchable silicon all use fabrication schemes that adhere the entirety of the silicon structure to a PDMS substrate. The robust adhesion results from condensation reactions to form covalent Si–O–Si bonds between a layer of SiO_2 on the silicon nanoribbons and hydroxyl groups on an oxidized PDMS surface [44]. Refining this approach by confining the adhesion to specific regions produces non-coplanar structures that distribute stress more effectively than planar structures, leading to systems that can withstand both complex and extreme mechanical deformations [45]. For example, patterning the adhesion of an ultrathin silicon mesh so that it is selectively anchored only at the intersection points results in the formation of connecting ribbons that lift upwards from the PDMS surface when the prestrain is released

(Figure 10a) [44, 46]. The resulting arc-shaped structures bend downwards with tensile strain, enabling the structure to reach elongations ~17 %. More importantly, the free-standing nature of these structures allows them to tolerate twisting deformations that are not available to fully bonded silicon nanoribbons (Figure 10b). As with planar structures, introducing serpentine-shaped structures to noncoplanar nanomeshes increases stretchability through changes in height combined with changes in the geometry of the serpentine shape (Figure 10c) [44]. Twisting and shearing of the serpentine bridges accommodate complex distributions of strain, enabling elongations of up to 140 % strain (Figure 10d).

The noncoplanar structures just described are based on the simple out-of-plane motion of a 2D mesh layout due to compressive buckling. Engineering the pattern of the layout and positioning of the adhesive regions enables the fabrication of more complex 3D structures [47–50]. For example, compressive buckling of serpentine structures with patterned adhesion to the elastomeric substrate results in the formation of a 3D helical structure similar to the classic coil spring (Figure 11). The stretching mechanics of these structures, also similar to those of coil springs, provide nearly ideal mechanics as helical interconnects in soft electronics.

1.3 Kirigami structures

Sections 1.1 and 1.2 detail how the field of stretchable electronics has developed an extraordinary array of design components based on the bending of wavy structures. Kirigami structures, inspired by the Japanese art of paper cutting and folding, have emerged as a distinctly different approach [51–54]. In kirigami structures, a pattern of periodic cuts in a solid membrane introduces controlled points of deformation, enabling the whole structure to accommodate stretching (Figure 12a) [52]. Depending on the kirigami pattern, rotation, bending, and twisting motions at designated local sites allow for global deformation and shape adaptability. By distributing the applied load throughout the kirigami structure, the function of the membrane can remain almost strain-invariant (Figure 12b) [52]. One interesting design involves cutting patterns that divide the planar membrane into connected units, which undergo in-plane rotations with stretching (Figure 12c) [55]. These designs can accommodate large omnidirectional strains and can be improved even further by applying fractal concepts [55, 56]. Just as fractal structures improved the stretchability of wavy structures by building in multiple levels of unraveling with strain, subdividing kirigami units by repeating the cut pattern within the original one creates a hierarchical structure that sequentially expands with stretching.

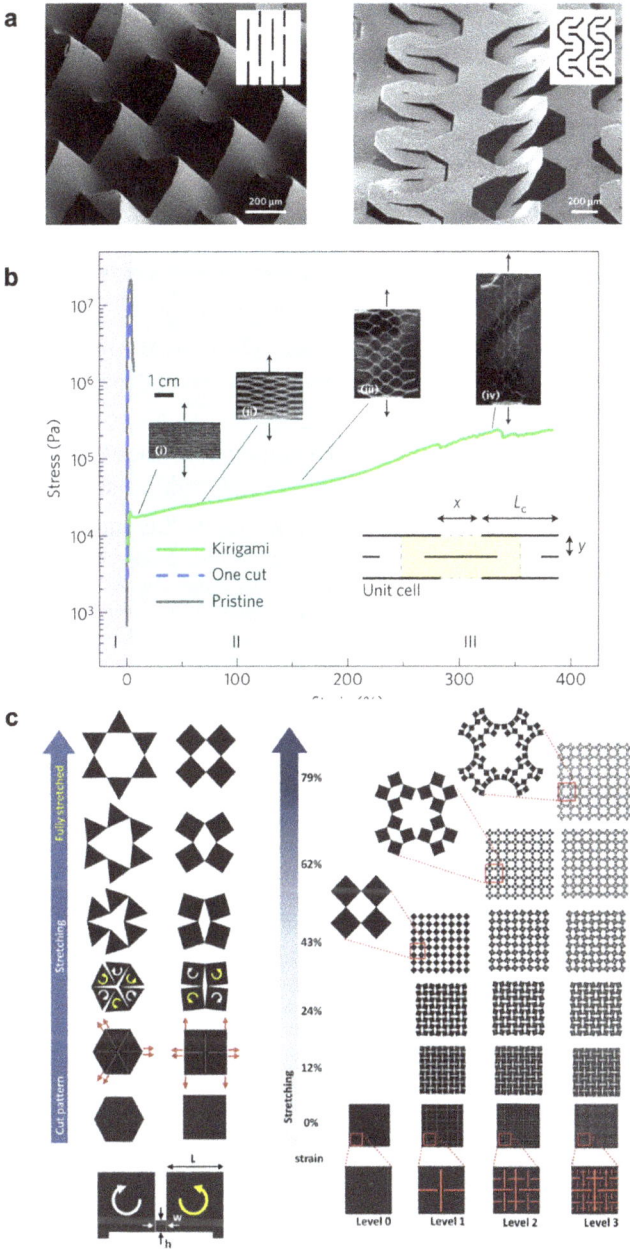

Figure 12: Kirigami structures. (a) Examples of kirigami structures. The insets show the corresponding kirigami unit cells. (b) Stress–strain curve for a model macroscale kirigami sheet (green), unpatterned sheet (grey curve), and a sheet with a single notch in the middle (dashed blue), with inset SEM images showing the deformed kirigami pattern at different strains. (c) Fractal kiragami design that separates the material into rotating units (left) and combinations of shapes into a multilevel hierarchy (right). Reproduced with permission from (a, b) reference [53], (c) reference [56].

1.4 Integration of design elements into device systems

The structures described in Sections 1.1–1.3 provide an assortment of design elements for stretchable electronics. Each of these design elements has specific properties and advantages that can be deployed to satisfy different device requirements and performance criteria. Furthermore, these designs are applicable to a range of functional materials, including metals, dielectrics, and even device-grade silicon. The importance of stretchable crystalline silicon cannot be overstated, as it enables the direct fabrication of high-performance, stretchable silicon-based electronic devices, eliminating the need to integrate rigid device islands and stretchable interconnects in the island–bridge architecture. Adding conventional fabrication steps that define patterns of dopants in the silicon nanoribbons and pattern metal contacts and interconnects and dielectric layers make it possible to fabricate devices such as stretchable p-n diodes, Schottky-barrier metal oxide semiconductor field-effect transistors (MOSFETs), and silicon photodetectors, as well as circuits such as CMOS logic gates, ring oscillators, and differential amplifiers [41, 57–59].

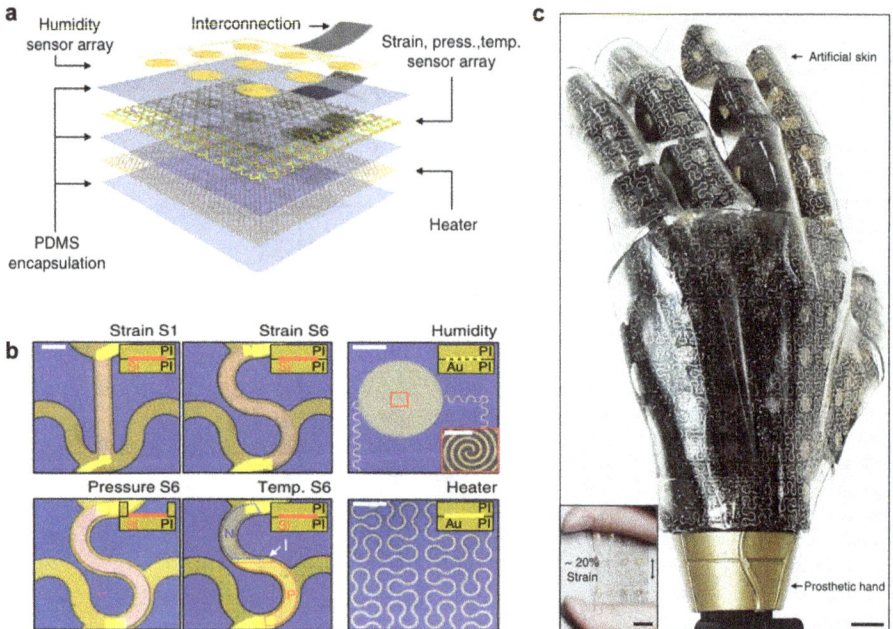

Figure 13: Smart prosthetic skin with multimodal sensing capabilities. (a) Exploded view of the artificial skin design, comprising six stacked layers. (b) Microscope images of strain, humidity, pressure, and temperature sensors configured into serpentine and fractal-inspired shapes that are tailored to the biomechanics of the prosthetic hand. (c) Photograph of the smart artificial skin with integrated stretchable sensors and actuators covering the entire surface area of a prosthetic hand. Scale bar, 1 cm. The inset shows the artificial skin stretched to 20 % strain. Scale bar, 1 cm. Reproduced with permission from reference [61].

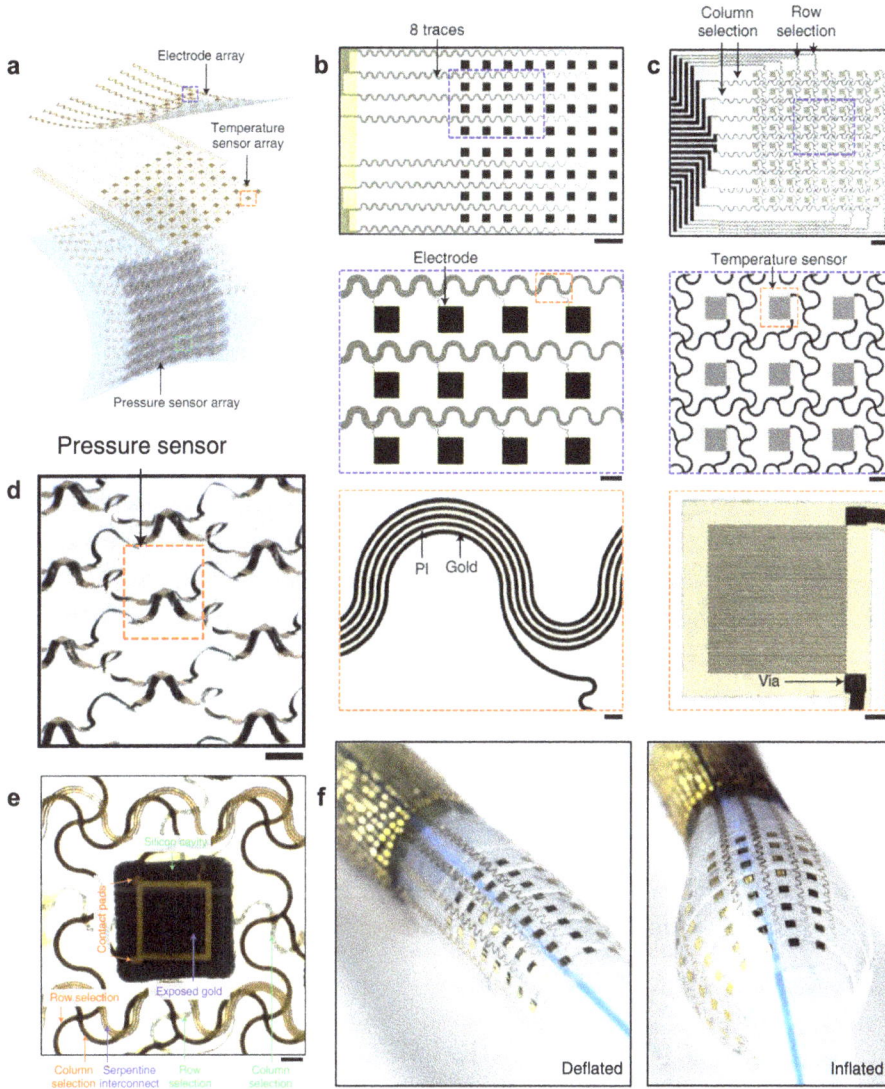

Figure 14: Deformable balloon catheters with combined diagnostic and therapeutic functions for minimally invasive surgery. (a) Schematic illustration of the multilayered device arrays. (b, c) Detailed images of the array of electrodes (b) and array of temperature sensors (c) in planar format showing the use of serpentines with redundant pathways and mesh layouts, respectively. Scale bars are 2 mm for the images in black outline, 500 μm for images in blue dashed outline, and 100 μm for images in orange dashed outline. (d) Images of the tent-like 3D strain gauges with 3D helical interconnects. Scale bar, 500 μm. (e) Image of one unit of the stacked arrays showing the layers for electrodes (blue), temperature sensors (orange), and pressure sensors (green). Scale bar, 200 μm. (f) Images of the array of electrodes transferred onto a balloon catheter of silicone. Scale bar, 2 mm. Reproduced with permission from reference [62].

The ingenious integration of selected device elements has produced some truly amazing device systems, which also highlights the maturity of this area of stretchable electronics research. For example, the smart prosthetic skin with multimodal sensing capabilities in Figure 13 integrates strain, pressure, and humidity sensors fabricated using p-type doped single crystalline silicon nanoribbons configured into serpentine shapes that are tailored to the mechanics of the prosthetic hand [60]. The system also includes an electroresistive heater in a fractal-inspired pattern to facilitate uniform heating. Via-hole structures interconnect the various circuit elements between the stacked layers. The resulting system holds great promise for smart prostheses endowed with the softness and sensory characteristics of human skin. In another example, incorporating stretchable structures into deformable balloon catheters used in cardiac surgery creates an advanced surgical tool that provides combined diagnostic and therapeutic functions (Figure 14) [61]. This multilayered design consists of an 8×8 array of sensors and actuators. Design elements include an array of gold electrodes connected by serpentine interconnects for electrophysiological recording and electrical stimulation, stretchable gold traces configured as resistive elements for precision thermography, and noncoplanar gold structures as pressure sensors. The pressure sensors result from the innovative use of patterned adhesion and compressive buckling to convert cross-shaped metal structures and serpentine interconnects into tent-like 3D strain gauges that are connected by 3D helical interconnects. These structures have a high sensitivity to normal force and a low effective modulus. Overall, the 8×8 pressure sensor array includes more than 400 bonding sites, more than 300 3D structures, and 64 3D pressure sensors. The resulting smart catheter addresses the modulus mismatch between conventional rigid devices and cardiac tissues, empowering surgeons to map temperature, pressure, and electrophysiological parameters, as well as perform localized functions such as radiofrequency ablation and/or irreversible electroporation at soft-tissue surfaces.

2 Functional materials that stretch

Rubbery materials can be stretched without breaking and then return to their original shape. Endowing electrically functional materials with such intrinsic stretchability forms an approach to stretchable electronics that is distinct from the architectural designs discussed in Section 1. Instead of configuring ultrathin materials into intentionally designed, sophisticated wavy shapes to absorb strain through nondestructive bending, achieving intrinsic stretchability focuses on engineering the mechanical properties of the materials themselves. The selection, synthesis, and/or modification of functional materials brings seemingly endless possibilities, spurring rapid growth in this dynamic area of research. While conveying the many ongoing developments in this field is a tall task, a good way to understand the field is to categorize it into three general approaches:

1) materials selection (liquid metals), 2) physical blending (stretchable composites), and 3) molecular/polymer engineering (molecularly stretchable materials).

2.1 Liquid metals

Liquids flow and change shape readily, making them naturally well-suited to stretchable electronics. Eutectic alloys of gallium and indium (EGaIn) and sometimes tin (GaInSn, galinstan), which are conductive liquids at room temperature, have dominated the liquid-based approach to stretchable electronics [62–66]. These alloys have a low elastic modulus, making them highly deformable; at the same time, they resist spontaneously adapting to form the lowest interfacial free energy shape due to the formation of a gallium oxide layer on the surface, which allows molding into metastable, nonspherical structures. In 2008, the Whitesides research group introduced EGaIn as a moldable soft electrode for the electrical characterization of self-assembled monolayer molecular junctions (Figure 15a) [67]. A subsequent investigation provided the field with a fundamental understanding of the rheological behavior of liquid metals (LMs) and furthermore demonstrated the great potential of incorporating LMs into PDMS microfluidic channels for stretchable electronics [68]. Since then, the field of LM microfluidic devices has flourished and has been reviewed extensively [69–71]. LMs contained within PDMS microchannels exhibit a low, even negligible, resistance change during stretching making them useful in a variety of devices, such as circuit interconnects, mechanically

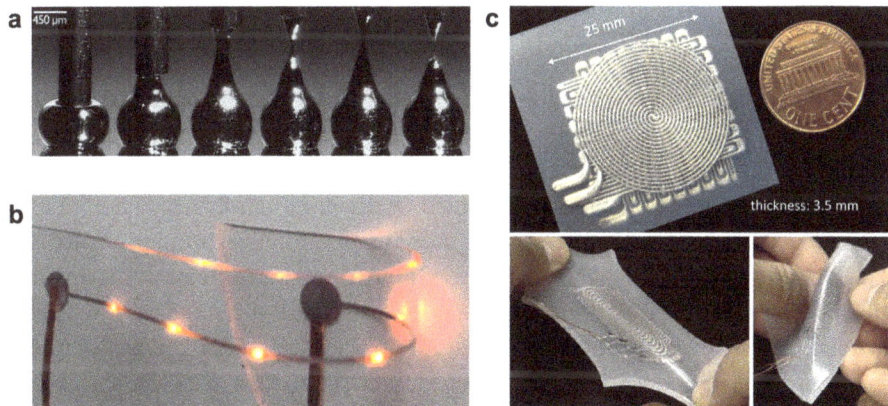

Figure 15: Liquid metal microfluidics. (a) A series of photographs showing the deformation of a drop of EGaIn dispensed from a syringe needle. As the needle is raised, the EGaIn deforms into a pear shape due to the formation of the oxide layer. (b) Multilayered strain/pressure sensor consisting of multilayered microchannels in an elastomer matrix filled with EGaIn. (c) Photograph of five LEDs interconnected with PDMS microfluidic channels filled with Galinstan. Reproduced with permission from (a) reference [67], (b) reference [73], (c) reference [72].

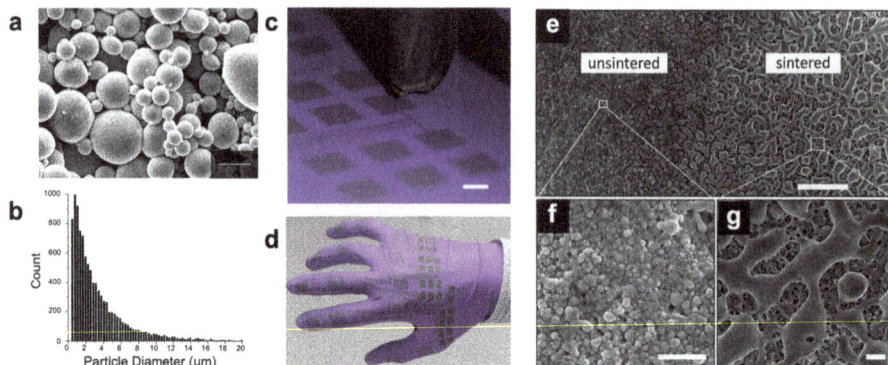

Figure 16: Formation and printing of liquid-metal particles. (a) SEM of EGaIn microparticles formed by ultrasonication (scale: 10 μm). (b) Distribution of EGaIn microparticle diameters after 30 s of ultrasonication. (c) Photograph of inkjet system printing EGaIn nanoparticles. Scale bar, 5 mm. d) Human hand wearing nitrile glove with inkjet-printed arrays of strain gauges, intricate wiring, and contact pads of EGaIn nanoparticles. (e) SEM images of a laser-sintered eGaIn nanoparticle film showing sintered-to-unsintered transition region across the edge of the pattern. Scale bar, 10 μm. (f, g) High magnification SEM images of the uncoalesced eGaIn nanoparticles (f) and coalesced eGaIn nanoparticle film after laser sintering (g). Reproduced with permission from (a, b) reference [84], (c, d) reference [82], (e–g) reference [86].

tunable antennas, wearable sensors, and energy harvesting, among others (Figure 15b, c) [66, 72–77].

The field of LM-based stretchable electronics significantly broadened through the fabrication of LM particles, which unlocked a readily accessible set of LM-based design elements by circumventing the need to fabricate PDMS microchannels [64, 78]. The oxide layer that forms on the surface of LMs enables them to be sheared into particles [64]. During shearing, the LM particles spontaneously form the outer oxide layer, stabilizing the particles and preventing them from re-forming their original bulk state. The size of the LM particles is controlled by the shearing method (e. g., simple manual agitation [79], microfluidic flow focusing [80], strong shear forces [81], and ultrasonication [82–84]) (Figure 16a, b). LM particles can then be dispersed in solvents and used as a printing "ink" to create conductive patterns on elastomers by inkjet printing (Figure 16c, d) [82] or spray coating [85]. In contrast, it is extremely challenging to push LMs in their bulk state through a printing nozzle due to the extremely high surface tension and presence of the native oxide. However, as-printed LM particles do not form conductive traces due to the presence of the oxide shell, which must be ruptured to allow the particles to merge. This merging can be achieved through laser sintering (Figure 16e–g) [86, 87] or simply stretching the elastomeric substrate, which provides enough mechanical stress to achieve electrical conductivity in a process known as "mechanical sintering" [82].

Some form of encapsulation is a requisite for any LM-based device. One way to use this requirement as an advantage involves dispersing the LM particles into elastomers like PDMS [62, 65]. Mixing LM particles with elastomers and curing yields liquid metal

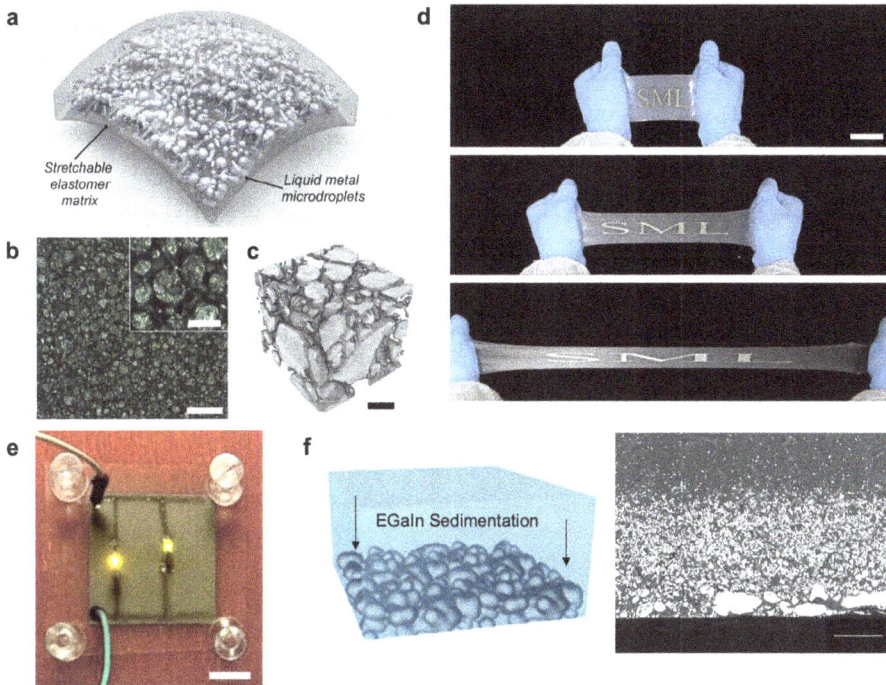

Figure 17: Liquid-metal embedded elastomers. (a) Schematic illustration showing the dispersion of liquid-metal drops in an elastomeric matrix. (b) Optical microscope images of the surface of LMEE composite. Scale bar, 100 μm and inset 25 μm. (c) Nano-CT scan showing the 3D microstructure of the LMEE. Scale bar, 25 μm. (d) Photographs demonstrating the stretchability of LMEE from 0 % (top) to 250 % (middle) to 500 % (bottom) strain. Scale bar, 5 cm. (e) A square sheet of a PDMS–galinstan LMEE with conductive lines drawn by selectively applying compression (dark regions). LEDs embedded into the conductive lines light up. (f) Schematic illustration and cross-sectional SEM of EGaIn microparticle sedimentation in PDMS to form iLMEE. d) of the EGaIn-rich layer of iLMEE (scale: 200 μm). Reproduced with permission from (a–d) reference [88], (e) reference [79], (f) reference [84].

embedded elastomers (LMEEs) that exhibit high thermal conductivity (Figure 17a–c) [88, 89]. These composites can achieve electrical conductivity through mechanical sintering, which ruptures the oxide shells and elastomeric membranes between LM particles. Furthermore, selectively applying mechanical pressure with a tip or using laser sintering enables the "drawing" of conductive circuit lines (Figure 17e) [79]. Building on this approach, allowing the LM particles to sediment during curing of the elastomer concentrates the particles at the bottom surface, reducing the amount of LM needed to achieve conductivity (Figure 17f) [84, 90–92].

A feature of all LM-based stretchable electronics is that the resistance remains almost unchanged to hundreds of percent strain, making them useful for a myriad of applications, from stretchable interconnects in island–bridge architectures to more sophisticated uses in soft robotics, wearable computing, and human-machine interfacing

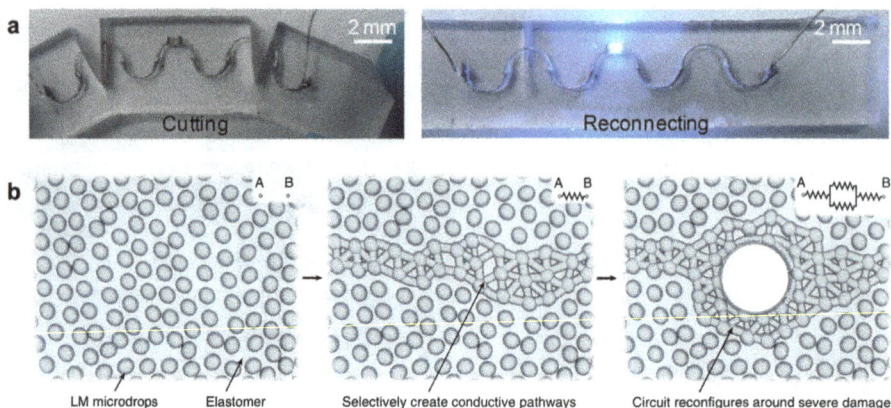

Figure 18: Self-healing liquid-metal electronics. (a) Photograph of a circuit comprising an LED connect to Galinstan-filled microchannels that has been physically separated with a blade (left) and reconnected (right). (b) Schematic illustration of LMEE (left) with selectively compression to create electrically conductive LM traces (middle), and the autonomous reconfiguring of the LMEE in response to damage (right). Reproduced with permission from (a) reference [93], (b) reference [94].

[63, 65]. Furthermore, LM electronics can be electrically self-healing: After cutting an LMEE composite or a microchannel filled with LM, electrical conductivity can be restored simply by bringing the interfaces back together (Figure 18a) [93]. Self-healing can also be achieved autonomously by including embedded, isolated LM droplets in the design. These drops act as autonomous self-healing agents by spontaneously rupturing to form new electrical connections in response to damage (Figure 18b) [94].

2.2 Stretchable composites

A composite material comprises two or more individual materials, each with different desirable properties. A straightforward approach to intrinsically stretchable functional materials involves combining functional materials with elastomers to form stretchable composites. The key to this strategy is maintaining interconnected pathways through the functional material that persist with stretching [95, 96]. A classic example is loading an elastomer with a carbon-black filler to form a conductive rubber; however, the electrical conductivity is too low to be useful in stretchable electronics [11]. In 2008, Sekitani et al. updated the formulation from carbon black to single-walled carbon nanotubes (SWNTs) to form highly conductive stretchable composites, demonstrating the importance of materials selection and blending method in the fabrication of functional composites. Dispersing SWCNTs into an ionic liquid by grinding forms a conductive "bucky gel" that is subsequently blended with a fluorinated copolymer [97]. The high aspect ratio of the SWNTs causes tangling, which preserves conductive pathways for charge transport to strains >100 %. The utility of this elastic conductor as a stretchable interconnect was

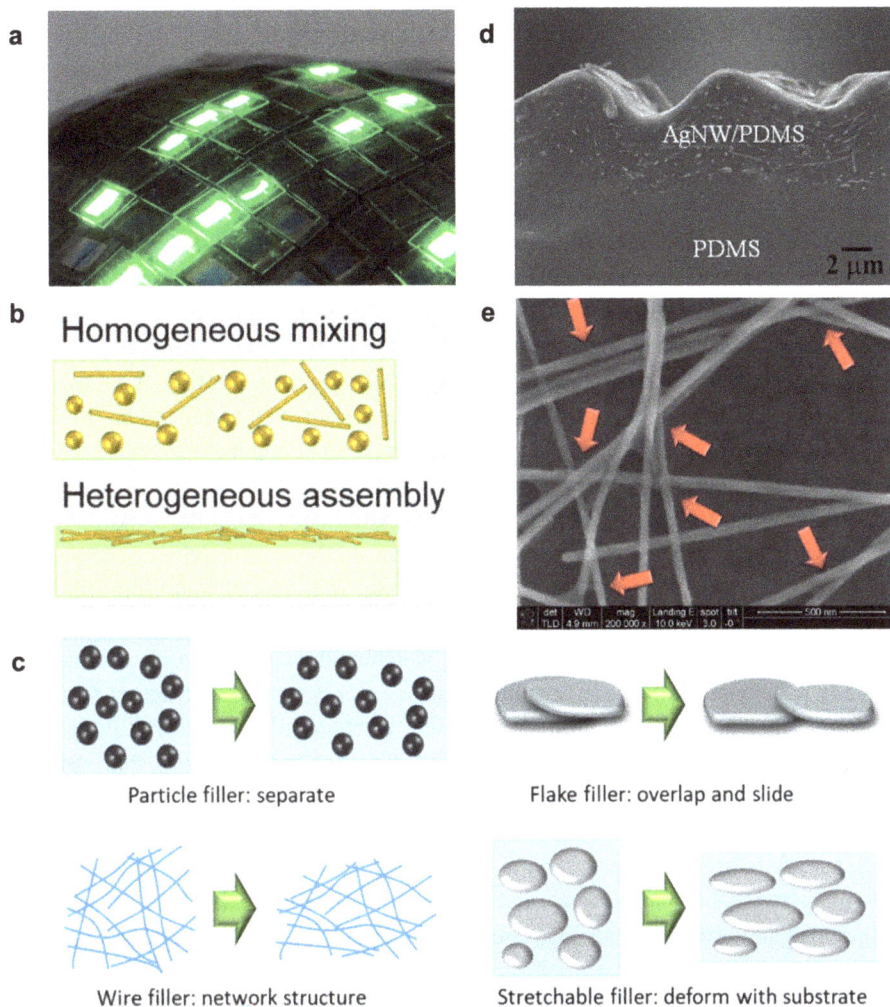

Figure 19: Stretchable composite materials. (a) Stretchable active-matrix organic light-emitting diode display using an island–bridge design with stretchable interconnects of carbon nanotubes dispersed in a fluorinated rubber. (b) Schematic illustration of two types of stretchable composites. (c) The strain response of stretchable composites with various filler shapes. (d) Wavy layered composite of silver nanowires on a PDMS substrate. (e) SEM image of stretchable silver nanowire network with graphene oxide soldered around the nanowire junctions (indicated by red arrows). Reproduced with permission from (a) reference [98], (b, c) reference [95], (d) reference [106], (e) reference [109].

demonstrated in a stretchable active-matrix organic light-emitting diode display using an island–bridge design (Figure 19a). The display functioned with stretching to 30–50 % elongation and also with stretching over a hemispherical surface [98].

Many other materials have been used to form stretchable composite conductors. These materials can be mixed with an elastomeric prepolymer and cured, or deposited or embedded at the surface of an elastomeric substrate (Figure 19b) [95]. Properties of the filler material that are important to consider include morphology (size, shape, and aspect ratio), electronic properties (conductivity), and mechanical properties (rigid or soft). Metal fillers such as metallic nanoparticles, nanoflakes, nanowires, and liquid metals illustrate the full spectrum of the effects of morphology (Figure 19c). Nanoparticles typically require high loading to reach the percolation threshold, which is the minimum loading of the nanomaterial required to form the conductive network [99]. Due to their low aspect ratio, stretching disconnects conductive pathways in the composite, leading to an increase in resistance [100]. The plate-like shape of metal flakes means they can slide past each other with stretching, exhibiting low sensitivity of the resistance to strain [101]. The high aspect ratio of metallic nanowires leads to the formation of a mesh-like structure that deforms with strain, with changes in resistance that depend on the loading of the nanowires [102, 103]. As discussed in Section 2.1, liquid metal particles deform with stretching, leading to almost no change in resistance with stretching. Allotropes of carbon (carbon black, graphene, graphene oxide, and carbon nanotubes) exhibit morphological effects similar to those of metals. However, the conductivity of composites that use carbon-based fillers is generally much lower than that of composites that use metallic fillers. At the same time, these classes of fillers differ in the mechanical properties they impart to stretchable composites. Metallic fillers (other than liquid metals) typically have poor mechanical flexibility, which reduces the tensile fracture strain of the composite. Carbon-based additives are more mechanically flexible, but increasing the filler content to achieve high conductivity stiffens the composite and thus reduces the fracture strain.

Although achieving high conductivity and mechanical stretchability together is a challenge, there are several ways to meet both requirements. Simply using deformable filler materials such as liquid metals and graphene is one approach [95, 104]. Graphene is a 2D carbon nanomaterial with good conductivity, high strength and toughness, and the ability to bend and deform within the elastic matrix [105]. Another simple and effective approach combines the architectural designs described in Section 1, such as serpentine shapes or wavy configurations that arise from prestraining the elastomer, with intrinsically stretchable composite materials. For example, wavy films of silver nanowires retain stable conductivity to 50 % elongation (Figure 19d) [106]. Combining different functional materials together in an elastomeric matrix is a way to take advantage of the properties of each material, such as combining fillers of different morphologies. For example, a blend of microscale silver flakes and silver nanoparticles produces a highly conductive stretchable composite [107]. The high conductivity comes from the silver flake, while the silver nanoparticles form a stable percolation network that persists with strain to preserve conductivity to strains as high as 400 %. A similar approach using carbon-based fillers combines CNTs with conductive CB particles [108]. In another example, combining AgNWs with graphene oxide (GO) forms a conductive network of AgNWs in which the GO sheets wrap around and solder the AgNW junctions to increase the stretchability of the composite (Figure 19e) [109]. The possibilities are endless.

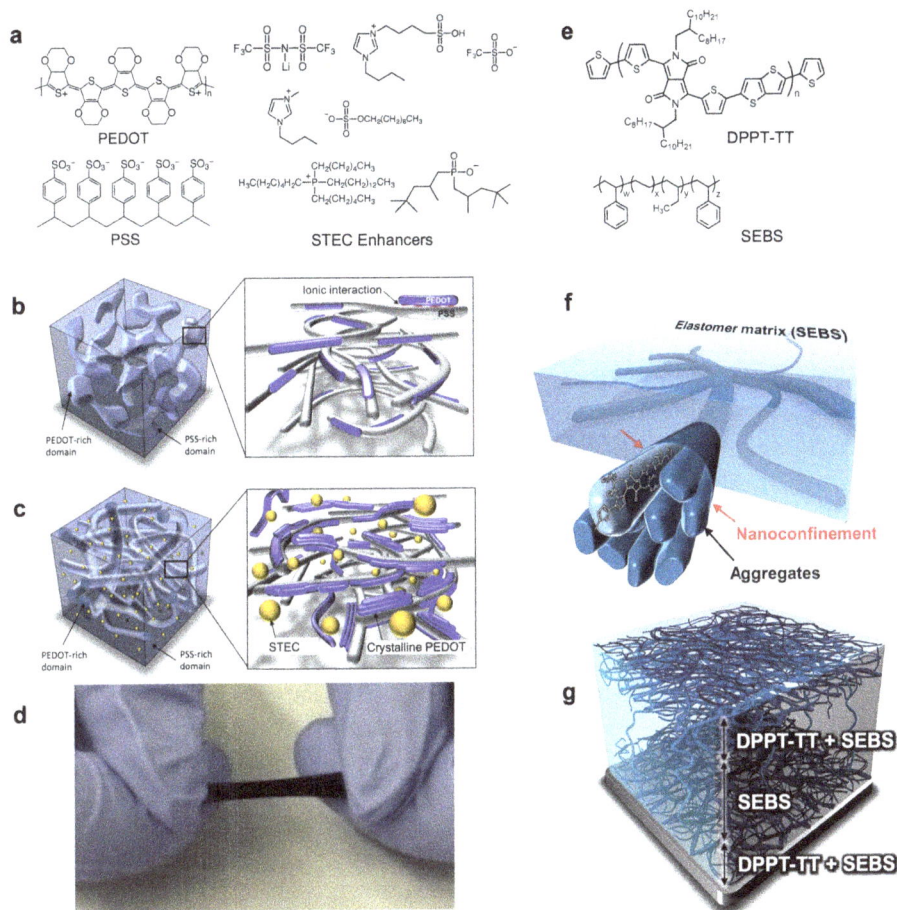

Figure 20: Stretchable polymer composites. (a) Chemical structure of PEDOT:PSS. (b) Schematic illustration of the morphology of a typical PEDOT:PSS film. (c) Schematic illustration of the morphology of that a stretchable PEDOT film with STEC enhancers. (d) Photograph of stretchable PEDOT:PSS/STEC composite. (e) Chemical structures of the semiconducting polymer DPPT-TT and SEBS elastomer. (f) Schematic illustration of the nanoconfinement effect. (g) Schematic illustration of the morphology of the nanoconfined DPPT-TT/SEBS composite. Reproduced with permission from (a–d) reference [111], (e–g) reference [119].

Conducting and semiconducting polymers are a large class of functional materials that are an important part of stretchable organic electronics. These materials are mostly π-conjugated polymers that consist of semiflexible polymer chains and are not intrinsically stretchable. As an example, PEDOT:PSS is one of the most widely used conducting polymers [110]. It is commercially available, highly stable, and transparent, and it can achieve a range of conductivities. However, it can function with stretching to only ~5 % elongation due to its semicrystalline structure (Figure 20a, b). Adding a plasticizer molecule such as glycerol, Triton X-100, Zonyl, or ionic liquids to PEDOT:PSS decreases the interaction between polymer chains and increases the free volume. These dopants not only enhance the stretchability, they can also act as secondary dopants to make the blend more conductive than PEDOT:PSS alone. Doping PEDOT:PSS with water-soluble ionic liquids—termed stretchability and electrical conductivity (STEC) enhancers— produces films that can be deposited on the surface of the elastomer styrene ethylene butylene styrene (SEBS). The resulting composite exhibits high conductivity that persists to elongations as high as 800 % (Figure 20c, d) [111]. Using PEDOT:PSS as a filler within an elastomeric matrix is more challenging due to its immiscibility with hydrophobic elastomers. Adding Triton-X-100 to PEDOT:PSS increases its miscibility with PDMS to provide composites in which the PEDOT:PSS is uniformly dispersed in the elastomer [112]. The sheet resistance of composites made using 60 % PEDOT:PSS is 20 ohm/sq, with conductivity that persists to ~80 % strain. Encouraging phase separation to create an interpenetrating polymer network (IPN) of PEDOT:PSS and PDMS boosts this performance [113]. In this approach, conditions are carefully controlled to embed a continuous network of PEDOT:PSS within PDMS. These stretchable IPNs combine the high conductivity of the 3D PEDOT:PSS network with the softness of PDMS. For example, filling the voids of a 3D PEDOT:PSS aerogel with PDMS provides an IPN with a sheet resistance of 2.5 ohm/sq using only 7.1 % PEDOT:PSS. This composite remains conductive to ~40 % strain [114].

Semiconducting polymers, a key part of organic electroluminescent devices and thin-film transistors, can also achieve stretchability through the composite approach [96, 115, 116]. Similar to the PEDOT:PSS composites discussed above, simple mixing of semiconducting polymers with elastomers results in improved mechanical stretchability [117]; however, the insulating elastomer regions impede charge transport between semiconducting domains and thus reduce charge carrier mobilities [116]. Phase separation in the composite is the key to high stretchability and enhanced charge-transport properties. For example, an IPN of poly(3-hexylthiophene-2,5-diyl) (P3HT) and PDMS can be formed by controlling the interdiffusion of the two materials during film formation [118]. The P3HT network provides efficient charge carrier transport pathways, even at very low loadings of ~0.5 wt %, and can be stretched to 100 % while still retaining the high charge carrier mobility. In a related approach, Xu et al. discovered that certain conditions and proper materials selection could favor phase separation into a 1D nanoconfined interconnected network of polymer semiconductor nanofibrils embedded in the elastomer [119]. This approach, termed conjugated polymer/elastomer phase-separation-induced elasticity (CONPHINE), was first demonstrated using the semiconducting polymer poly(2,5-bis(2-octyldodecyl)-3,6-di(thiophen-2-yl)diketopyrrolo[3,4-c]pyrrole-1,4-dione-alt-thieno[3,2-b]thiophen) (DPPT-TT) and the thermoplastic elastomer SEBS (Figure 20e). SEBS induces the formation of DPPT-TT nanoaggregates in solution, which form 1D nanoconfined nanofibers in the elastomer matrix during thin film formation (Figure 20f, g). The nanoconfinement effect provides both improved mechanical properties and higher charge-carrier mobilities compared to those of the corresponding neat polymer semiconductor [120, 121]. CONPHINE films

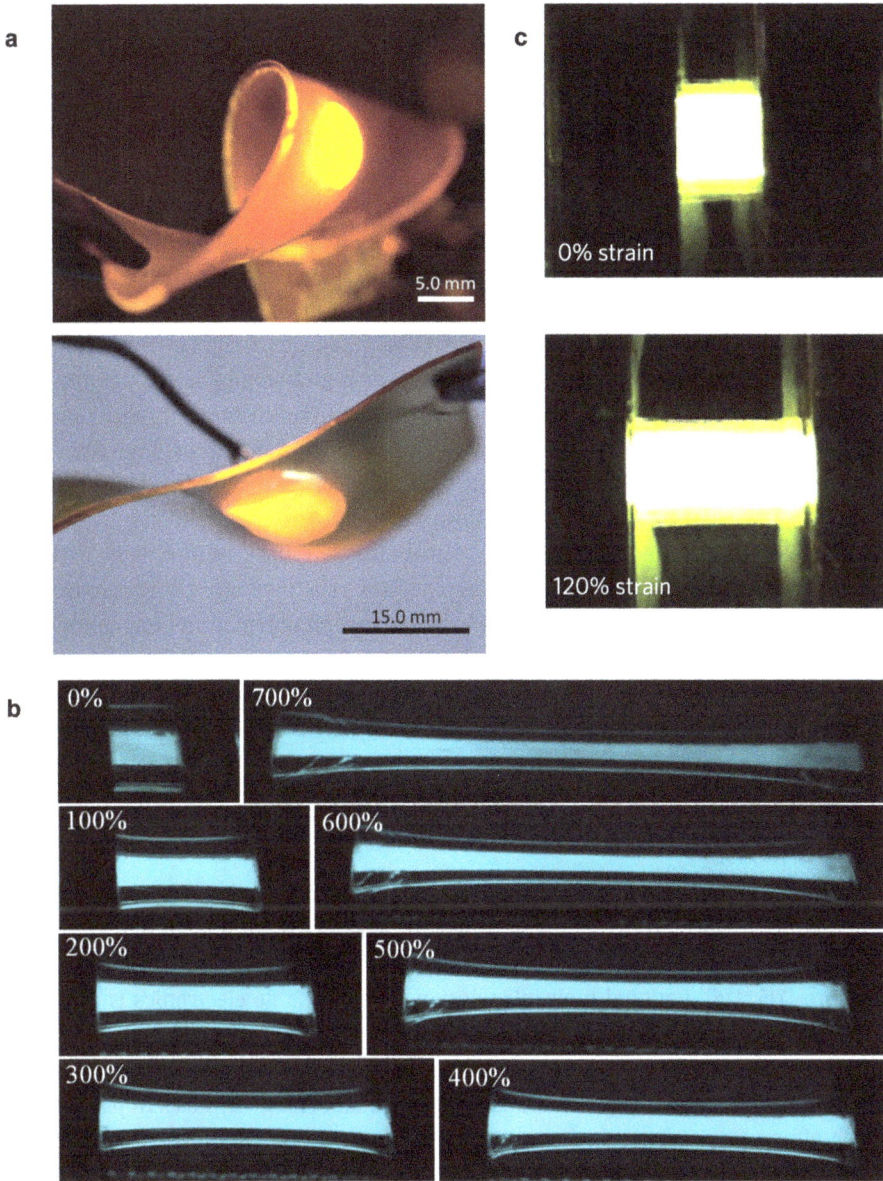

Figure 21: Stretchable electroluminescent materials. (a) Intrinsically stretchable light-emitting electrochemical cell fabricated using a composite of Ru(bpy)$_3$(PF$_6$)$_2$ and PDMS. (b) Extremely stretchable alternating current electroluminescent devices fabricated using a composite of copper-doped zinc sulfide particles and Ecoflex. (c) Stretchable polymer light-emitting electrochemical cell fabricated using the semiconducting polymer SuperYellow. Reproduced with permission from (a) reference [124], (b) reference [125], (c) reference [126].

of DPPT-TT and SEBS could be stretched to 100 % strain without affecting the charge carrier mobility.

Stretchable composites also make use of electroluminescent materials as fillers to form intrinsically stretchable thin-film devices that provide uniform, large-area light emission that persists during stretching [122, 123]. Electroluminescent materials can include molecular complexes, particles, and polymers. These devices are fabricated by sandwiching a layer of the stretchable emissive composite between two intrinsically stretchable electrodes. For example, Filiatrault et al. created a stretchable light-emitting electrochemical cell (LEC) by combining an emissive material—an ionic ruthenium complex—with the elastomer PDMS (Figure 21a) [124]. Similarly, Lee et al. reported blending ZnS:Cu electroluminescent particles with the elastomer Ecoflex to produce alternating current electroluminescent (ACEL) devices that functioned to high mechanical strains of 700 % (Figure 21b) [125]. Polymeric emissive materials have also been used in polymer LECs (PLECs). PLECs made using an emissive composite of the semiconducting polymer SuperYellow with an ionic mixture of ethoxylated trimethylolpropane, poly(ethylene oxide), and lithium trifluoromethane sulfonate emitted light to 120 % strain, but exhibited limited resilience to cycles of repetitive stretching (Figure 21c) [126]. However, configuring the emissive semiconducting polymer and ion-conducting mixture into an IPN provided rubbery elasticity and light emission that persisted to elongations as large as 140 %, as well as stable performance with repeated stretching at 50 % strain for at least 1,000 cycles [127].

2.3 Molecularly stretchable materials

Rather than relying on additives like plasticizers or blending with elastomers, the development of molecularly stretchable materials seeks to use rational design to synthesize functional materials that are intrinsically stretchable by virtue of their molecular structure [115, 116]. Rational molecular design for stretchable electronics is a rapidly growing area of research with vast potential to create new materials that simultaneously achieve elasticity and high electrical performance and maintain efficient charge-transport pathways over repeated strain cycles to enable long-term use. At present, research efforts focus on producing a variety of new materials to develop insight into fundamental molecular structure–morphology–property relationships to establish the design rules for molecularly stretchable materials. Much of this work has focused on molecularly stretchable semiconducting polymers.

Polymer semiconductors are π-conjugated systems that transport-charge carriers along the conjugated polymer backbone (intrachain transport) and between different polymer chains through π–π stacking (interchain transport) (Figure 22a) [128]. It is also the π–π stacking that causes the formation of rigid crystalline domains within thin films of polymer semiconductors that are prone to fracturing under strain and thus reduce the stretchability. Maximizing both the charge carrier mobility and stretchability—

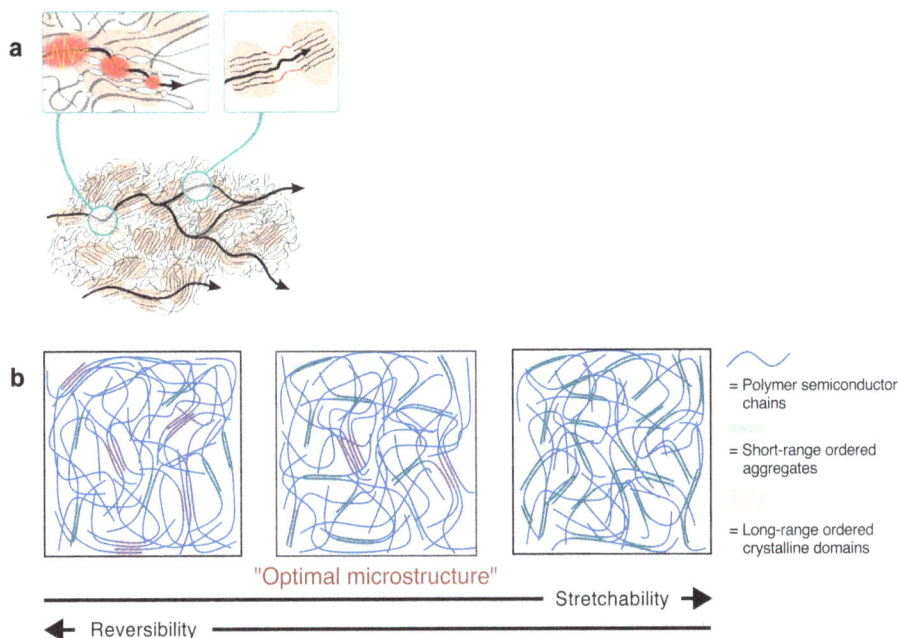

Figure 22: Properties of polymer semiconductors. (a) Schematic illustration of charge transport in conjugated semiconducting polymers. Polymer chains are indicated by grey lines, the orange shaded regions indicate the semicrystalline domains of the polymer, and the red shaded regions shows electron wavefunctions with varying degrees of localization. (b) Range of microstructures of polymer semiconductors, showing the "optimal microstructure" that supports high charge-carrier mobility, stretchability, and mechanical reversibility. Reproduced with permission from (a) reference [128], (b) reference [129].

properties that seem at odds with each other—requires detailed polymer engineering to control the fraction and size distribution of crystalline domains within softer amorphous regions to achieve the "ideal microstructure" shown in Figure 22b. Careful control of these crystalline domains can also provide the important property of reversible stretchability by acting as physical crosslinking sites. Here, we summarize the general approaches to polymer engineering that continue to evolve in this rapidly growing field.

Intrinsic stretchability depends on the ability of the polymer to dissipate strain. The incorporation of multiple strain energy-dissipation mechanisms into the molecular design enables the material to absorb and dissipate strain energy during mechanical deformations, preventing irreversible damage to the material during stretching and preserving charge-transport pathways (Figure 23a) [116]. One way polymers dissipate strain is through conformational changes, which prompted the development of conjugated semiconducting polymers with easy access to conformational changes in response to mechanical stress. Copolymerizing a fraction of nonconjugated spacers into the polymer backbone introduces amorphous regions that provide the desired conformational flexibility (Figure 23b) [129]. However, breaking the conjugation of the backbone dimin-

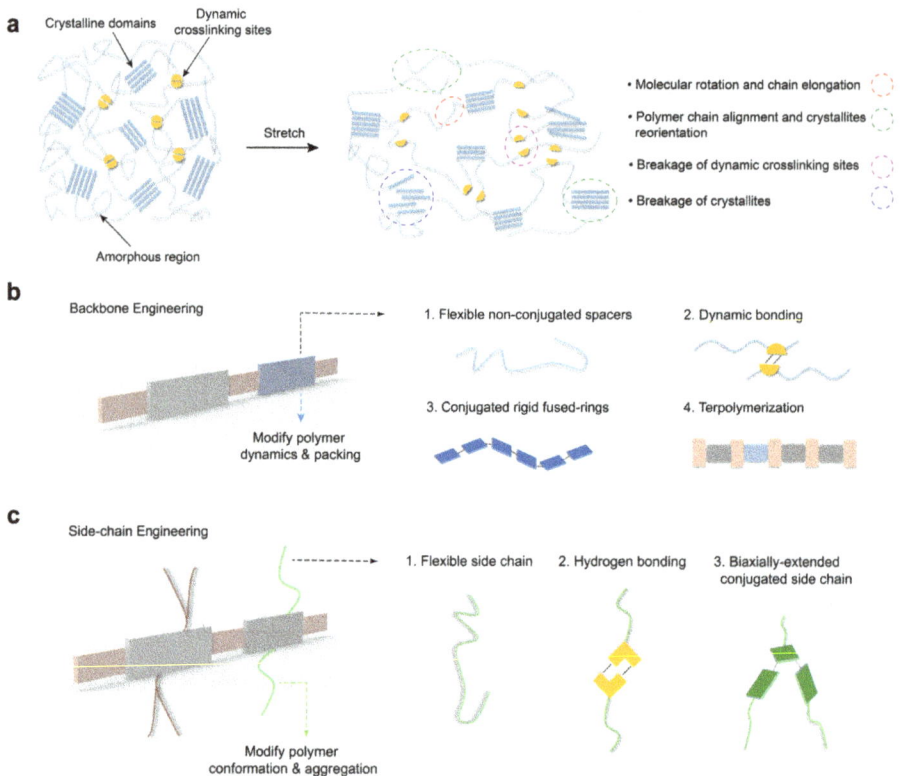

Figure 23: Polymer engineering. (a) Schematic illustration of the mechanisms to dissipate strain energy in a polymer semiconductor thin film during stretching. Each mechanism is illustrated by a dashed circle of a different color. (b) Schematic illustration of the design principles for backbone engineering. (c) Schematic illustration of the design principles for side-chain engineering. Reproduced with permission from reference [116].

ishes intrachain charge transport and weakens the interchain packing. The result is a disruption of the long-range crystalline order, which can decrease the charge-carrier mobility. The trick is to find the optimum proportion of nonconjugated spacers that balance stretchability and electrical performance. It is also possible to introduce conjugated rigid fused rings into the polymer backbone rather than nonconjugated spacers to preserve charge-carrier mobility (Figure 23b) [130]. Of course, this approach diminishes the stretchability by reducing the conformational flexibility of the backbone. Elaborating on the polymer design by incorporating side chains can improve the mechanical properties by affecting the polymer chain conformation, aggregation in solution, and packing in the solid state (Figure 23c) [116]. For example, nonconjugated side chains can alter the molecular packing in a way that reduces the elastic modulus of the film [131]. Properties of nonconjugated side chains like chain length and branching are important variables to consider in the molecular design.

Dynamic bonding is another effective strain dissipation mechanism. Incorporating noncovalent crosslinking groups in the polymer design introduces connections between polymer chains that reversibly break and re-form in response to strain. Hydrogen bonds [132] and metal-ligand coordination bonds [133] can be incorporated in the backbone or on the side chains to dissipate strain and provide mechanical resilience. The strength of the dynamic bonds and their number and position in the polymer structure are important parameters. In particular, the great diversity of metal-ligand interactions, including the type and oxidation state of the metal center, the strength of the metal-ligand interaction, the denticity of the coordinating ligands, and the type of counter ion, have the potential to provide exquisite control over the strength of non-covalent interactions that determine the mechanical properties of the polymer. Dynamic bonding can also impart self-healing properties to conjugated polymers [134]. The vision for self-healing soft electronics is inspired by the ability of biological systems, such as human skin, to spontaneously repair physical damage. Dynamic bonds can be used as sacrificial breakage sites in response to strain that rapidly reconnect upon release of the strain, thus healing the materials.

3 Outlook

In 2005, More than Moore posed the question: *What devices do we really want to wear?* The field of stretchable electronics has responded over the past ~20 years with an extensive array of design elements to enable the fabrication of electronic devices that are soft, stretchable, twistable, and bendable. The integration of these device elements has provided exciting new devices that conform to human skin, with innumerable applications in human health and wellness, safety, soft robotics, and more. Although it is impossible to capture this diverse and exciting field in one chapter, the overview presented here tells the story of how this field has evolved and how our understanding of architectural designs and intrinsically stretchable materials has deepened through the creative work of countless researchers. In the ensuing chapters, researchers who continue to advance stretchable electronics describe the continuing evolution of the field—from new opportunities to leverage architectural designs to the ongoing advances in the rational design of new materials.

Bibliography

[1]   Graef, M. More than moore white paper. In *2021 IEEE international roadmap for devices and systems outbriefs*; IEEE: Santa Clara, CA, USA, 2021; pp. 1–47. https://doi.org/10.1109/IRDS54852.2021.00013.
[2]   Lacour, S. P.; Wagner, S.; Huang, Z.; Suo, Z. Stretchable gold conductors on elastomeric substrates. *Appl. Phys. Lett.* **2003**, *82* (15), 2404–2406. https://doi.org/10.1063/1.1565683.

[3]   Savagatrup, S.; Printz, A. D.; O'Connor, T. F.; Zaretski, A. V.; Lipomi, D. J. Molecularly stretchable electronics. *Chem. Mater.* **2014**, *26* (10), 3028–3041. https://doi.org/10.1021/cm501021v.

[4]   Mechael, S. S.; Wu, Y.; Schlingman, K.; Carmichael, T. B. Stretchable metal films. *Flex. Print. Electron.* **2018**, *3* (4), 043001. https://doi.org/10.1088/2058-8585/aae8c0.

[5]   Filiatrault, H. L.; Carmichael, R. S.; Boutette, R. A.; Carmichael, T. B. A self-assembled, low-cost, microstructured layer for extremely stretchable gold films. *ACS Appl. Mater. Interfaces* **2015**, *7* (37), 20745–20752. https://doi.org/10.1021/acsami.5b05677.

[6]   Wu, Y.; Schlingman, K.; Mechael, S. S.; Chen, Y.; Carmichael, T. B. Engineering the cracking patterns in stretchable copper films using acid-oxidized poly(dimethylsiloxane) substrates. *ACS Appl. Electron. Mater.* **2022**, *4* (11), 5565–5572. https://doi.org/10.1021/acsaelm.2c01161.

[7]   Mandlik, P.; Lacour, S. P.; Li, J. W.; Chou, S. Y.; Wagner, S. Fully elastic interconnects on nanopatterned elastomeric substrates. *IEEE Electron Device Lett.* **2006**, *27* (8), 3.

[8]   Mechael, S. S.; D'Amaral, G. M.; Wu, Y.; Schlingman, K.; Ives, B.; Carmichael, R. S.; Carmichael, T. B. The synergistic effect of topography and stiffness as a crack engineering strategy for stretchable electronics. *J. Mater. Chem. C* **2023**, *11* (2), 497–512. https://doi.org/10.1039/D2TC03459J.

[9]   Chen, Y.; Wu, Y.; Mechael, S. S.; Carmichael, T. B. Heterogeneous surface orientation of solution-deposited gold films enables retention of conductivity with high strain—A new strategy for stretchable electronics. *Chem. Mater.* **2019**, *31* (6), 1920–1927. https://doi.org/10.1021/acs.chemmater.8b04487.

[10] Amjadi, M.; Kyung, K.; Park, I.; Sitti, M. Stretchable, skin-mountable, and wearable strain sensors and their potential applications: a review. *Adv. Funct. Mater.* **2016**, *26* (11), 1678–1698. https://doi.org/10.1002/adfm.201504755.

[11] Rogers, J. A.; Someya, T.; Huang, Y. Materials and mechanics for stretchable electronics. *Science* **2010**, *327* (5973), 1603–1607. https://doi.org/10.1126/science.1182383.

[12] Xue, Z.; Song, H.; Rogers, J. A.; Zhang, Y.; Huang, Y. Mechanically-guided structural designs in stretchable inorganic electronics. *Adv. Mater.* **2020**, *32* (15), 1902254. https://doi.org/10.1002/adma.201902254.

[13] Lacour, S. P.; Jones, J.; Wagner, S.; Li, T.; Suo, Z. Stretchable interconnects for elastic electronic surfaces. *Proc. IEEE* **2005**, *93* (8), 1459–1467. https://doi.org/10.1109/JPROC.2005.851502.

[14] Gray, D. S.; Tien, J.; Chen, C. S. High-conductivity elastomeric electronics. *Adv. Mater.* **2004**, *16* (5), 393–397. https://doi.org/10.1002/adma.200306107.

[15] Li, T.; Suo, Z.; Lacour, S. P.; Wagner, S. Compliant thin film patterns of stiff materials as platforms for stretchable electronics. *J. Mater. Res.* **2005**, *20* (12), 3274–3277. https://doi.org/10.1557/jmr.2005.0422.

[16] Zhou, C.; Bette, S.; Schnakenberg, U. Flexible and stretchable gold microstructures on extra soft poly(dimethylsiloxane) substrates. *Adv. Mater.* **2015**, *27* (42), 6664–6669. https://doi.org/10.1002/adma.201502630.

[17] Widlund, T.; Yang, S.; Hsu, Y.-Y.; Lu, N. Stretchability and compliance of freestanding serpentine-shaped ribbons. *Int. J. Solids Struct.* **2014**, *51* (23–24), 4026–4037. https://doi.org/10.1016/j.ijsolstr.2014.07.025.

[18] Zhang, Y.; Xu, S.; Fu, H.; Lee, J.; Su, J.; Hwang, K.-C.; Rogers, J. A.; Huang, Y. Buckling in serpentine microstructures and applications in elastomer-supported ultra-stretchable electronics with high areal coverage. *Soft Matter* **2013**, *9* (33), 8062. https://doi.org/10.1039/c3sm51360b.

[19] Fan, Z.; Zhang, Y.; Ma, Q.; Zhang, F.; Fu, H.; Hwang, K.-C.; Huang, Y. A finite deformation model of planar serpentine interconnects for stretchable electronics. *Int. J. Solids Struct.* **2016**, *91*, 46–54. https://doi.org/10.1016/j.ijsolstr.2016.04.030.

[20] Brosteaux, D.; Axisa, F.; Gonzalez, M.; Vanfleteren, J. Design and fabrication of elastic interconnections for stretchable electronic circuits. *IEEE Electron Device Lett.* **2007**, *28* (7), 552–554. https://doi.org/10.1109/LED.2007.897887.

[21] Jahanshahi, A.; Salvo, P.; Vanfleteren, J. Reliable stretchable gold interconnects in biocompatible elastomers. *J. Polym. Sci., Part B, Polym. Phys.* **2012**, *50* (11), 773–776. https://doi.org/10.1002/polb.23064.

[22] Gonzalez, M.; Axisa, F.; Vanden Bulcke, M.; Brosteaux, D.; Vandevelde, B.; Vanfleteren, J. Design of metal interconnects for stretchable electronic circuits using finite element analysis. In *2007 international conference on thermal, mechanical and multi-physics simulation experiments in microelectronics and micro-systems. EuroSime 2007*; IEEE: London, 2007; pp. 1–6. https://doi.org/10.1109/ESIME.2007.360005.

[23] Qaiser, N.; Damdam, A. N.; Khan, S. M.; Bunaiyan, S.; Hussain, M. M. Design criteria for horseshoe and spiral-based interconnects for highly stretchable electronic devices. *Adv. Funct. Mater.* **2021**, *31* (7), 2007445. https://doi.org/10.1002/adfm.202007445.

[24] Jahanshahi, A.; Gonzalez, M.; Brand, J. V. D.; Bossuyt, F.; Vervust, T.; Verplancke, R.; Vanfleteren, J.; Baets, J. D. Stretchable circuits with horseshoe shaped conductors embedded in elastic polymers. *Jpn. J. Appl. Phys.* **2013**, *52* (5S1), 05DA18. https://doi.org/10.7567/JJAP.52.05DA18.

[25] Lu, N.; Yang, S. Mechanics for stretchable sensors. *Curr. Opin. Solid State Mater. Sci.* **2015**, *19* (3), 149–159. https://doi.org/10.1016/j.cossms.2014.12.007.

[26] Zhang, Y.; Fu, H.; Su, Y.; Xu, S.; Cheng, H.; Fan, J. A.; Hwang, K.-C.; Rogers, J. A.; Huang, Y. Mechanics of ultra-stretchable self-similar serpentine interconnects. *Acta Mater.* **2013**, *61* (20), 7816–7827. https://doi.org/10.1016/j.actamat.2013.09.020.

[27] Xu, S.; Zhang, Y.; Cho, J.; Lee, J.; Huang, X.; Jia, L.; Fan, J. A.; Su, Y.; Su, J.; Zhang, H.; Cheng, H.; Lu, B.; Yu, C.; Chuang, C.; Kim, T.; Song, T.; Shigeta, K.; Kang, S.; Dagdeviren, C.; Petrov, I.; Braun, P. V.; Huang, Y.; Paik, U.; Rogers, J. A. Stretchable batteries with self-similar serpentine interconnects and integrated wireless recharging systems. *Nat. Commun.* **2013**, *4* (1), 1543. https://doi.org/10.1038/ncomms2553.

[28] Dong, W.; Zhu, C.; Ye, D.; Huang, Y. Optimal design of self-similar serpentine interconnects embedded in stretchable electronics. *Appl. Phys. A* **2017**, *123* (6), 428. https://doi.org/10.1007/s00339-017-0985-3.

[29] Fan, J. A.; Yeo, W.-H.; Su, Y.; Hattori, Y.; Lee, W.; Jung, S.-Y.; Zhang, Y.; Liu, Z.; Cheng, H.; Falgout, L.; Bajema, M.; Coleman, T.; Gregoire, D.; Larsen, R. J.; Huang, Y.; Rogers, J. A. Fractal design concepts for stretchable electronics. *Nat. Commun.* **2014**, *5* (1), 3266. https://doi.org/10.1038/ncomms4266.

[30] Xu, L.; Gutbrod, S. R.; Ma, Y.; Petrossians, A.; Liu, Y.; Webb, R. C.; Fan, J. A.; Yang, Z.; Xu, R.; Whalen, J. J.; Weiland, J. D.; Huang, Y.; Efimov, I. R.; Rogers, J. A. Materials and fractal designs for 3D multifunctional integumentary membranes with capabilities in cardiac electrotherapy. *Adv. Mater.* **2015**, *27* (10), 1731–1737. https://doi.org/10.1002/adma.201405017.

[31] Norton, J. J. S.; Lee, D. S.; Lee, J. W.; Lee, W.; Kwon, O.; Won, P.; Jung, S.-Y.; Cheng, H.; Jeong, J.-W.; Akce, A.; Umunna, S.; Na, I.; Kwon, Y. H.; Wang, X.-Q.; Liu, Z.; Paik, U.; Huang, Y.; Bretl, T.; Yeo, W.-H.; Rogers, J. A. Soft, curved electrode systems capable of integration on the auricle as a persistent brain–computer interface. *Proc. Natl. Acad. Sci.* **2015**, *112* (13), 3920–3925. https://doi.org/10.1073/pnas.1424875112.

[32] Sung, W.-L.; Chen, C.-C.; Huang, K.; Fang, W. Development of a large-area chip network with multidevice integration using a stretchable electroplated copper spring. *J. Micromech. Microeng.* **2016**, *26* (2), 025003. https://doi.org/10.1088/0960-1317/26/2/025003.

[33] Lv, C.; Yu, H.; Jiang, H. Archimedean spiral design for extremely stretchable interconnects. *Extrem. Mech. Lett.* **2014**, *1*, 29–34. https://doi.org/10.1016/j.eml.2014.12.008.

[34] Rojas, J. P.; Arevalo, A.; Foulds, I. G.; Hussain, M. M. Design and characterization of ultra-stretchable monolithic silicon fabric. *Appl. Phys. Lett.* **2014**, *105* (15), 154101. https://doi.org/10.1063/1.4898128.

[35] Volynskii, A. L.; Bazhenov, S.; Lebedeva, O. V.; Bakeev, N. F. Mechanical buckling instability of thin coatings deposited on soft polymer substrates.

[36] Lacour, S. P.; Jones, J.; Suo, Z.; Wagner, S. Design and performance of thin metal film interconnects for skin-like electronic circuits. *IEEE Electron Device Lett.* **2004**, *25* (4), 179–181. https://doi.org/10.1109/LED.2004.825190.

[37] Rodríguez-Hernández, J. Wrinkled interfaces: taking advantage of surface instabilities to pattern polymer surfaces. *Prog. Polym. Sci.* **2015**, *42*, 1–41. https://doi.org/10.1016/j.progpolymsci.2014.07.008.

[38] Ma, T.; Liang, H.; Chen, G.; Poon, B.; Jiang, H.; Yu, H. Micro-strain sensing using wrinkled stiff thin films on soft substrates as tunable optical grating. *Opt. Express* **2013**, *21* (10), 11994. https://doi.org/10.1364/OE.21.011994.

[39] Xu, Y.; Chen, M.; Yu, S.; Zhou, H. High-performance flexible strain sensors based on silver film wrinkles modulated by liquid pdms substrates. *RSC Adv.* **2023**, *13* (48), 33697–33706. https://doi.org/10.1039/D3RA06020A.

[40] Choi, W. M.; Song, J.; Khang, D.-Y.; Jiang, H.; Huang, Y. Y.; Rogers, J. A. Biaxially stretchable "wavy" silicon nanomembranes. *Nano Lett.* **2007**, *7* (6), 1655–1663. https://doi.org/10.1021/nl0706244.

[41] Khang, D.-Y.; Jiang, H.; Huang, Y.; Rogers, J. A. A stretchable form of single-crystal silicon for high-performance electronics on rubber substrates. *Science* **2006**, *311* (5758), 208–212. https://doi.org/10.1126/science.1121401.

[42] Jiang, H.; Khang, D.-Y.; Song, J.; Sun, Y.; Huang, Y.; Rogers, J. A. Finite deformation mechanics in buckled thin films on compliant supports. *Proc. Natl. Acad. Sci.* **2007**, *104* (40), 15607–15612. https://doi.org/10.1073/pnas.0702927104.

[43] Sim, K.; Li, Y.; Song, J.; Yu, C. Biaxially stretchable ultrathin si enabled by serpentine structures on prestrained elastomers. *Adv. Mater. Technol.* **2019**, *4* (1), 1800489. https://doi.org/10.1002/admt.201800489.

[44] Kim, D.-H.; Song, J.; Choi, W. M.; Kim, H.-S.; Kim, R.-H.; Liu, Z.; Huang, Y. Y.; Hwang, K.-C.; Zhang, Y.; Rogers, J. A. Materials and noncoplanar mesh designs for integrated circuits with linear elastic responses to extreme mechanical deformations. *Proc. Natl. Acad. Sci.* **2008**, *105* (48), 18675–18680. https://doi.org/10.1073/pnas.0807476105.

[45] Song, J.; Huang, Y.; Xiao, J.; Wang, S.; Hwang, K. C.; Ko, H. C.; Kim, D.-H.; Stoykovich, M. P.; Rogers, J. A. Mechanics of noncoplanar mesh design for stretchable electronic circuits. *J. Appl. Phys.* **2009**, *105* (12), 123516. https://doi.org/10.1063/1.3148245.

[46] Ko, H. C.; Shin, G.; Wang, S.; Stoykovich, M. P.; Lee, J. W.; Kim, D.; Ha, J. S.; Huang, Y.; Hwang, K.; Rogers, J. A. Curvilinear electronics formed using silicon membrane circuits and elastomeric transfer elements. *Small* **2009**, *5* (23), 2703–2709. https://doi.org/10.1002/smll.200900934.

[47] Jang, K.-I.; Li, K.; Chung, H. U.; Xu, S.; Jung, H. N.; Yang, Y.; Kwak, J. W.; Jung, H. H.; Song, J.; Yang, C.; Wang, A.; Liu, Z.; Lee, J. Y.; Kim, B. H.; Kim, J.-H.; Lee, J.; Yu, Y.; Kim, B. J.; Jang, H.; Yu, K. J.; Kim, J.; Lee, J. W.; Jeong, J.-W.; Song, Y. M.; Huang, Y.; Zhang, Y.; Rogers, J. A. Self-assembled three dimensional network designs for soft electronics. *Nat. Commun.* **2017**, *8* (1), 15894. https://doi.org/10.1038/ncomms15894.

[48] Xu, S.; Yan, Z.; Jang, K.-I.; Huang, W.; Fu, H.; Kim, J.; Wei, Z.; Flavin, M.; McCracken, J.; Wang, R.; Badea, A.; Liu, Y.; Xiao, D.; Zhou, G.; Lee, J.; Chung, H. U.; Cheng, H.; Ren, W.; Banks, A.; Li, X.; Paik, U.; Nuzzo, R. G.; Huang, Y.; Zhang, Y.; Rogers, J. A. Assembly of micro/nanomaterials into complex, three-dimensional architectures by compressive buckling. *Science* **2015**, *347* (6218), 154–159. https://doi.org/10.1126/science.1260960.

[49] Liu, Y.; Yan, Z.; Lin, Q.; Guo, X.; Han, M.; Nan, K.; Hwang, K.; Huang, Y.; Zhang, Y.; Rogers, J. A. Guided formation of 3D helical mesostructures by mechanical buckling: analytical modeling and experimental validation. *Adv. Funct. Mater.* **2016**, *26* (17), 2909–2918. https://doi.org/10.1002/adfm.201505132.

[50] Yan, Z.; Han, M.; Yang, Y.; Nan, K.; Luan, H.; Luo, Y.; Zhang, Y.; Huang, Y.; Rogers, J. A. Deterministic assembly of 3D mesostructures in advanced materials via compressive buckling: a short review of recent progress. *Extrem. Mech. Lett.* **2017**, *11*, 96–104. https://doi.org/10.1016/j.eml.2016.12.006.

[51] Tao, J.; Khosravi, H.; Deshpande, V.; Li, S. Engineering by cuts: how kirigami principle enables unique mechanical properties and functionalities. *Adv. Sci.* **2023**, *10* (1), 2204733. https://doi.org/10.1002/advs.202204733.

[52] Shyu, T. C.; Damasceno, P. F.; Dodd, P. M.; Lamoureux, A.; Xu, L.; Shlian, M.; Shtein, M.; Glotzer, S. C.; Kotov, N. A. A kirigami approach to engineering elasticity in nanocomposites through patterned defects. *Nat. Mater.* **2015**, *14* (8), 785–789. https://doi.org/10.1038/nmat4327.

[53] Zhang, Y.; Yan, Z.; Nan, K.; Xiao, D.; Liu, Y.; Luan, H.; Fu, H.; Wang, X.; Yang, Q.; Wang, J.; Ren, W.; Si, H.; Liu, F.; Yang, L.; Li, H.; Wang, J.; Guo, X.; Luo, H.; Wang, L.; Huang, Y.; Rogers, J. A. A mechanically driven form of kirigami as a route to 3D mesostructures in micro/nanomembranes. *Proc. Natl. Acad. Sci.* **2015**, *112* (38), 11757–11764. https://doi.org/10.1073/pnas.1515602112.

[54] Song, Z.; Wang, X.; Lv, C.; An, Y.; Liang, M.; Ma, T.; He, D.; Zheng, Y.-J.; Huang, S.-Q.; Yu, H.; Jiang, H. Kirigami-based stretchable Lithium-ion batteries. *Sci. Rep.* **2015**, *5* (1), 10988. https://doi.org/10.1038/srep10988.

[55] Cho, Y.; Shin, J.-H.; Costa, A.; Kim, T. A.; Kunin, V.; Li, J.; Lee, S. Y.; Yang, S.; Han, H. N.; Choi, I.-S.; Srolovitz, D. J. Engineering the shape and structure of materials by fractal cut. *Proc. Natl. Acad. Sci.* **2014**, *111* (49), 17390–17395. https://doi.org/10.1073/pnas.1417276111.

[56] Tang, Y.; Yin, J. Design of cut unit geometry in hierarchical kirigami-based auxetic metamaterials for high stretchability and compressibility. *Extrem. Mech. Lett.* **2017**, *12*, 77–85. https://doi.org/10.1016/j.eml.2016.07.005.

[57] Kim, D.-H.; Lu, N.; Ma, R.; Kim, Y.-S.; Kim, R.-H.; Wang, S.; Wu, J.; Won, S. M.; Tao, H.; Islam, A.; Yu, K. J.; Kim, T.; Chowdhury, R.; Ying, M.; Xu, L.; Li, M.; Chung, H.-J.; Keum, H.; McCormick, M.; Liu, P.; Zhang, Y.-W.; Omenetto, F. G.; Huang, Y.; Coleman, T.; Rogers, J. A. Epidermal electronics. *Science* **2011**, *333*, 838–843.

[58] Kim, D.-H.; Ahn, J.-H.; Choi, W. M.; Kim, H.-S.; Kim, T.-H.; Song, J.; Huang, Y. Y.; Liu, Z.; Lu, C.; Rogers, J. A. Stretchable and foldable silicon integrated circuits. *Science* **2008**, *320* (5875), 507–511. https://doi.org/10.1126/science.1154367.

[59] Kim, B. H.; Lee, J.; Won, S. M.; Xie, Z.; Chang, J.-K.; Yu, Y.; Cho, Y. K.; Jang, H.; Jeong, J. Y.; Lee, Y.; Ryu, A.; Kim, D. H.; Lee, K. H.; Lee, J. Y.; Liu, F.; Wang, X.; Huo, Q.; Min, S.; Wu, D.; Ji, B.; Banks, A.; Kim, J.; Oh, N.; Jin, H. M.; Han, S.; Kang, D.; Lee, C. H.; Song, Y. M.; Zhang, Y.; Huang, Y.; Jang, K.-I.; Rogers, J. A. Three-dimensional silicon electronic systems fabricated by compressive buckling process. *ACS Nano* **2018**, *12* (5), 4164–4171. https://doi.org/10.1021/acsnano.8b00180.

[60] Kim, J.; Lee, M.; Shim, H. J.; Ghaffari, R.; Cho, H. R.; Son, D.; Jung, Y. H.; Soh, M.; Choi, C.; Jung, S.; Chu, K.; Jeon, D.; Lee, S.-T.; Kim, J. H.; Choi, S. H.; Hyeon, T.; Kim, D.-H. Stretchable silicon nanoribbon electronics for skin prosthesis. *Nat. Commun.* **2014**, *5* (1), 5747. https://doi.org/10.1038/ncomms6747.

[61] Han, M.; Chen, L.; Aras, K.; Liang, C.; Chen, X.; Zhao, H.; Li, K.; Faye, N. R.; Sun, B.; Kim, J.-H.; Bai, W.; Yang, Q.; Ma, Y.; Lu, W.; Song, E.; Baek, J. M.; Lee, Y.; Liu, C.; Model, J. B.; Yang, G.; Ghaffari, R.; Huang, Y.; Efimov, I. R.; Rogers, J. A. Catheter-integrated soft multilayer electronic arrays for multiplexed sensing and actuation during cardiac surgery. *Nat. Biomed. Eng.* **2020**, *4* (10), 997–1009. https://doi.org/10.1038/s41551-020-00604-w.

[62] Majidi, C.; Alizadeh, K.; Ohm, Y.; Silva, A.; Tavakoli, M. Liquid metal polymer composites: from printed stretchable circuits to soft actuators. *Flex. Print. Electron.* **2022**, *7* (1), 013002. https://doi.org/10.1088/2058-8585/ac515a.

[63] Dickey, M. D. Stretchable and soft electronics using liquid metals. *Adv. Mater.* **2017**, *29* (27), 1606425. https://doi.org/10.1002/adma.201606425.

[64] Lin, Y.; Genzer, J.; Dickey, M. D. Attributes, fabrication, and applications of gallium-based liquid metal particles. *Adv. Sci.* **2020**, *7* (12), 2000192. https://doi.org/10.1002/advs.202000192.

[65] Malakooti, M. H.; Bockstaller, M. R.; Matyjaszewski, K.; Majidi, C. Liquid metal nanocomposites. *Nanoscale Adv.* **2020**, *2* (7), 2668–2677. https://doi.org/10.1039/D0NA00148A.

[66] Won, P.; Jeong, S.; Majidi, C.; Ko, S. H. Recent advances in liquid-metal-based wearable electronics and materials. *iScience* **2021**, *24* (7), 102698. https://doi.org/10.1016/j.isci.2021.102698.

[67] Chiechi, R. C.; Weiss, E. A.; Dickey, M. D.; Whitesides, G. M. Eutectic gallium–indium (EGaIn): a moldable liquid metal for electrical characterization of self-assembled monolayers. *Angew. Chem., Int. Ed.* **2008**, *47* (1), 142–144. https://doi.org/10.1002/anie.200703642.

[68] Dickey, M. D.; Chiechi, R. C.; Larsen, R. J.; Weiss, E. A.; Weitz, D. A.; Whitesides, G. M. Eutectic gallium-indium (EGaIn): a liquid metal alloy for the formation of stable structures in microchannels at room temperature. *Adv. Funct. Mater.* **2008**, *18* (7), 1097–1104. https://doi.org/10.1002/adfm.200701216.

[69] Paracha, K. N.; Butt, A. D.; Alghamdi, A. S.; Babale, S. A.; Soh, P. J. Liquid metal antennas: materials, fabrication and applications. *Sensors* **2019**, *20* (1), 177. https://doi.org/10.3390/s20010177.

[70] Khoshmanesh, K.; Tang, S.-Y.; Zhu, J. Y.; Schaefer, S.; Mitchell, A.; Kalantar-Zadeh, K.; Dickey, M. D. Liquid metal enabled microfluidics. *Lab Chip* **2017**, *17* (6), 974–993. https://doi.org/10.1039/C7LC00046D.

[71] Zhu, L.; Wang, B.; Handschuh-Wang, S.; Zhou, X. Liquid metal–based soft microfluidics. *Small* **2020**, *16* (9), 1903841. https://doi.org/10.1002/smll.201903841.

[72] Park, Y.-L.; Chen, B.-R.; Wood, R. J. Design and fabrication of soft artificial skin using embedded microchannels and liquid conductors. *IEEE Sens. J.* **2012**, *12* (8), 2711–2718. https://doi.org/10.1109/JSEN.2012.2200790.

[73] Jeong, S. H.; Hagman, A.; Hjort, K.; Jobs, M.; Sundqvist, J.; Wu, Z. Liquid alloy printing of microfluidic stretchable electronics. *Lab Chip* **2012**, *12* (22), 4657. https://doi.org/10.1039/c2lc40628d.

[74] Baharfar, M.; Kalantar-Zadeh, K. Emerging role of liquid metals in sensing. *ACS Sens.* **2022**, *7* (2), 386–408. https://doi.org/10.1021/acssensors.1c02606.

[75] Dickey, M. D. Emerging applications of liquid metals featuring surface oxides. *ACS Appl. Mater. Interfaces* **2014**, *6* (21), 18369–18379. https://doi.org/10.1021/am5043017.

[76] So, J.; Thelen, J.; Qusba, A.; Hayes, G. J.; Lazzi, G.; Dickey, M. D. Reversibly deformable and mechanically tunable fluidic antennas. *Adv. Funct. Mater.* **2009**, *19* (22), 3632–3637. https://doi.org/10.1002/adfm.200900604.

[77] Zadan, M.; Chiew, C.; Majidi, C.; Malakooti, M. H. Liquid metal architectures for soft and wearable energy harvesting devices. *Multifunct. Mater.* **2021**, *4* (1), 012001. https://doi.org/10.1088/2399-7532/abd4f0.

[78] Chiew, C.; Morris, M. J.; Malakooti, M. H. Functional liquid metal nanoparticles: synthesis and applications. *Mater. Adv.* **2021**, *2* (24), 7799–7819. https://doi.org/10.1039/D1MA00789K.

[79] Fassler, A.; Majidi, C. Liquid-phase metal inclusions for a conductive polymer composite. *Adv. Mater.* **2015**, *27* (11), 1928–1932. https://doi.org/10.1002/adma.201405256.

[80] Thelen, J.; Dickey, M. D.; Ward, T. A study of the production and reversible stability of EGaIn liquid metal microspheres using flow focusing. *Lab Chip* **2012**, *12* (20), 3961. https://doi.org/10.1039/c2lc40492c.

[81] Tevis, I. D.; Newcomb, L. B.; Thuo, M. Synthesis of liquid core–shell particles and solid patchy multicomponent particles by shearing liquids into complex particles (SLICE). *Langmuir* **2014**, *30* (47), 14308–14313. https://doi.org/10.1021/la5035118.

[82] Boley, J. W.; White, E. L.; Kramer, R. K. Mechanically sintered gallium–indium nanoparticles. *Adv. Mater.* **2015**, *27* (14), 2355–2360. https://doi.org/10.1002/adma.201404790.

[83] Yamaguchi, A.; Mashima, Y.; Iyoda, T. Reversible size control of liquid-metal nanoparticles under ultrasonication. *Angew. Chem., Int. Ed.* **2015**, *54* (43), 12809–12813. https://doi.org/10.1002/anie.201506469.

[84] Schlingman, K.; D'Amaral, G. M.; Carmichael, R. S.; Carmichael, T. B. Intrinsically conductive liquid metal-elastomer composites for stretchable and flexible electronics. *Adv. Mater. Technol.* **2023**, *8* (1), 2200374. https://doi.org/10.1002/admt.202200374.

[85] Mohammed, M. G.; Kramer, R. All-printed flexible and stretchable electronics. *Adv. Mater.* **2017**, *29* (19), 1604965. https://doi.org/10.1002/adma.201604965.

[86] Liu, S.; Yuen, M. C.; White, E. L.; Boley, J. W.; Deng, B.; Cheng, G. J.; Kramer-Bottiglio, R. Laser sintering of liquid metal nanoparticles for scalable manufacturing of soft and flexible electronics. *ACS Appl. Mater. Interfaces* **2018**, *10* (33), 28232–28241. https://doi.org/10.1021/acsami.8b08722.

[87] Liu, S.; Reed, S. N.; Higgins, M. J.; Titus, M. S.; Kramer-Bottiglio, R. Oxide rupture-induced conductivity in liquid metal nanoparticles by laser and thermal sintering. *Nanoscale* **2019**, *11* (38), 17615–17629. https://doi.org/10.1039/C9NR03903A.

[88] Bartlett, M. D.; Fassler, A.; Kazem, N.; Markvicka, E. J.; Mandal, P.; Majidi, C. Stretchable, high-*k* dielectric elastomers through liquid-metal inclusions. *Adv. Mater.* **2016**, *28* (19), 3726–3731. https://doi.org/10.1002/adma.201506243.

[89] Bartlett, M. D.; Kazem, N.; Powell-Palm, M. J.; Huang, X.; Sun, W.; Malen, J. A.; Majidi, C. High thermal conductivity in soft elastomers with elongated liquid metal inclusions. *Proc. Natl. Acad. Sci.* **2017**, *114* (9), 2143–2148. https://doi.org/10.1073/pnas.1616377114.

[90] Yu, Z.; Shang, J.; Niu, X.; Liu, Y.; Liu, G.; Dhanapal, P.; Zheng, Y.; Yang, H.; Wu, Y.; Zhou, Y.; Wang, Y.; Tang, D.; Li, R. A composite elastic conductor with high dynamic stability based on 3D-calabash bunch conductive network structure for wearable devices. *Adv. Electron. Mater.* **2018**, *4* (9), 1800137. https://doi.org/10.1002/aelm.201800137.

[91] Ford, M. J.; Patel, D. K.; Pan, C.; Bergbreiter, S.; Majidi, C. Controlled assembly of liquid metal inclusions as a general approach for multifunctional composites. *Adv. Mater.* **2020**, *32* (46), 2002929. https://doi.org/10.1002/adma.202002929.

[92] Neumann, T. V.; Facchine, E. G.; Leonardo, B.; Khan, S.; Dickey, M. D. Direct write printing of a self-encapsulating liquid metal–silicone composite. *Soft Matter* **2020**, *16* (28), 6608–6618. https://doi.org/10.1039/D0SM00803F.

[93] Li, G.; Wu, X.; Lee, D.-W. A Galinstan-based inkjet printing system for highly stretchable electronics with self-healing capability. *Lab Chip* **2016**, *16* (8), 1366–1373. https://doi.org/10.1039/C6LC00046K.

[94] Markvicka, E. J.; Bartlett, M. D.; Huang, X.; Majidi, C. An autonomously electrically self-healing liquid metal–elastomer composite for robust soft-matter robotics and electronics. *Nat. Mater.* **2018**, *17* (7), 618–624. https://doi.org/10.1038/s41563-018-0084-7.

[95] Yun, G.; Tang, S.-Y.; Lu, H.; Zhang, S.; Dickey, M. D.; Li, W. Hybrid-filler stretchable conductive composites: from fabrication to application. *Small Sci.* **2021**, *1* (6), 2000080. https://doi.org/10.1002/smsc.202000080.

[96] Shim, H. J.; Sunwoo, S.; Kim, Y.; Koo, J. H.; Kim, D. Functionalized elastomers for intrinsically soft and biointegrated electronics. *Adv. Healthc. Mater.* **2021**, *10* (17), 2002105. https://doi.org/10.1002/adhm.202002105.

[97] Sekitani, T.; Noguchi, Y.; Hata, K.; Fukushima, T.; Aida, T.; Someya, T. A rubberlike stretchable active matrix using elastic conductors. *Science* **2008**, *321* (5895), 1468–1472. https://doi.org/10.1126/science.1160309.

[98] Sekitani, T.; Nakajima, H.; Maeda, H.; Fukushima, T.; Aida, T.; Hata, K.; Someya, T. Stretchable active-matrix organic light-emitting diode display using printable elastic conductors. *Nat. Mater.* **2009**, *8* (6), 494–499. https://doi.org/10.1038/nmat2459.

[99] Taherian, R. Development of an equation to model electrical conductivity of polymer-based carbon nanocomposites. *ECS J. Solid State Sci. Technol.* **2014**, *3* (6), M26–M38. https://doi.org/10.1149/2.023406jss.

[100] Feng, P.; Ye, Z.; Wang, Q.; Chen, Z.; Wang, G.; Liu, X.; Li, K.; Zhao, W. Stretchable and conductive composites film with efficient electromagnetic interference shielding and absorptivity. *J. Mater. Sci.* **2020**, *55* (20), 8576–8590. https://doi.org/10.1007/s10853-019-04172-6.

[101] Matsuhisa, N.; Kaltenbrunner, M.; Yokota, T.; Jinno, H.; Kuribara, K.; Sekitani, T.; Someya, T. Printable elastic conductors with a high conductivity for electronic textile applications. *Nat. Commun.* **2015**, *6* (1), 7461. https://doi.org/10.1038/ncomms8461.

[102] Lin, Y.; Li, Q.; Ding, C.; Wang, J.; Yuan, W.; Liu, Z.; Su, W.; Cui, Z. High-resolution and large-size stretchable electrodes based on patterned silver nanowires composites. *Nano Res.* **2022**, *15* (5), 4590–4598. https://doi.org/10.1007/s12274-022-4088-x.

[103] Chen, Y.; Carmichael, R. S.; Carmichael, T. B. Patterned, flexible, and stretchable silver nanowire/polymer composite films as transparent conductive electrodes. *ACS Appl. Mater. Interfaces* **2019**, *11* (34), 31210–31219. https://doi.org/10.1021/acsami.9b11149.

[104] Wang, J.; Cai, G.; Li, S.; Gao, D.; Xiong, J.; Lee, P. S. Printable superelastic conductors with extreme stretchability and robust cycling endurance enabled by liquid-metal particles. *Adv. Mater.* **2018**, *30* (16), 1706157. https://doi.org/10.1002/adma.201706157.

[105] Stankovich, S.; Dikin, D. A.; Dommett, G. H. B.; Kohlhaas, K. M.; Zimney, E. J.; Stach, E. A.; Piner, R. D.; Nguyen, S. T.; Ruoff, R. S. Graphene-based composite materials. *Nature* **2006**, *442* (7100), 282–286. https://doi.org/10.1038/nature04969.

[106] Xu, F.; Zhu, Y. Highly conductive and stretchable silver nanowire conductors. *Adv. Mater.* **2012**, *24* (37), 5117–5122. https://doi.org/10.1002/adma.201201886.

[107] Matsuhisa, N.; Inoue, D.; Zalar, P.; Jin, H.; Matsuba, Y.; Itoh, A.; Yokota, T.; Hashizume, D.; Someya, T. Printable elastic conductors by in situ formation of silver nanoparticles from silver flakes. *Nat. Mater.* **2017**, *16* (8), 834–840. https://doi.org/10.1038/nmat4904.

[108] Zheng, Y.; Li, Y.; Dai, K.; Wang, Y.; Zheng, G.; Liu, C.; Shen, C. A highly stretchable and stable strain sensor based on hybrid carbon nanofillers/polydimethylsiloxane conductive composites for large human motions monitoring. *Compos. Sci. Technol.* **2018**, *156*, 276–286. https://doi.org/10.1016/j.compscitech.2018.01.019.

[109] Liang, J.; Li, L.; Tong, K.; Ren, Z.; Hu, W.; Niu, X.; Chen, Y.; Pei, Q. Silver nanowire percolation network soldered with graphene oxide at room temperature and its application for fully stretchable polymer light-emitting diodes. *ACS Nano* **2014**, *8* (2), 1590–1600. https://doi.org/10.1021/nn405887k.

[110] Kayser, L. V.; Lipomi, D. J. Stretchable Conductive polymers and composites based on PEDOT and PEDOT:PSS. *Adv. Mater.* **2019**, *31* (10), 1806133. https://doi.org/10.1002/adma.201806133.

[111] Wang, Y.; Zhu, C.; Pfattner, R.; Yan, H.; Jin, L.; Chen, S.; Molina-Lopez, F.; Lissel, F.; Liu, J.; Rabiah, N. I.; Chen, Z.; Chung, J. W.; Linder, C.; Toney, M. F.; Murmann, B.; Bao, Z. A highly stretchable, transparent, and conductive polymer. *Sci. Adv.* **2017**, *3* (3), e1602076. https://doi.org/10.1126/sciadv.1602076.

[112] Luo, R.; Li, H.; Du, B.; Zhou, S.; Zhu, Y. A simple strategy for high stretchable, flexible and conductive polymer films based on PEDOT:PSS-PDMS blends. *Org. Electron.* **2020**, *76*, 105451. https://doi.org/10.1016/j.orgel.2019.105451.

[113] Yang, Y.; Deng, H.; Fu, Q. Recent progress on PEDOT:PSS based polymer blends and composites for flexible electronics and thermoelectric devices. *Mater. Chem. Front.* **2020**, *4* (11), 3130–3152. https://doi.org/10.1039/D0QM00308E.

[114] Teng, C.; Lu, X.; Zhu, Y.; Wan, M.; Jiang, L. Polymer in situ embedding for highly flexible, stretchable and water stable PEDOT:PSS composite conductors. *RSC Adv.* **2013**, *3* (20), 7219. https://doi.org/10.1039/c3ra41124a.

[115] Wang, G. N.; Gasperini, A.; Bao, Z. Stretchable polymer semiconductors for plastic electronics. *Adv. Electron. Mater.* **2018**, *4* (2), 1700429. https://doi.org/10.1002/aelm.201700429.

[116] Zheng, Y.; Zhang, S.; Tok, J. B.-H.; Bao, Z. Molecular design of stretchable polymer semiconductors: current progress and future directions. *J. Am. Chem. Soc.* **2022**, *144* (11), 4699–4715. https://doi.org/10.1021/jacs.2c00072.

[117] Carpi, F.; Gallone, G.; Galantini, F.; De Rossi, D. Silicone–poly(hexylthiophene) blends as elastomers with enhanced electromechanical transduction properties. *Adv. Funct. Mater.* **2008**, *18* (2), 235–241. https://doi.org/10.1002/adfm.200700757.

[118] Zhang, G.; McBride, M.; Persson, N.; Lee, S.; Dunn, T. J.; Toney, M. F.; Yuan, Z.; Kwon, Y.-H.; Chu, P.-H.; Risteen, B.; Reichmanis, E. Versatile interpenetrating polymer network approach to robust stretchable electronic devices. *Chem. Mater.* **2017**, *29* (18), 7645–7652. https://doi.org/10.1021/acs.chemmater.7b03019.

[119] Xu, J.; Wang, S.; Wang, G.-J. N.; Zhu, C.; Luo, S.; Jin, L.; Gu, X.; Chen, S.; Feig, V. R.; To, J. W. F.; Rondeau-Gagné, S.; Park, J.; Schroeder, B. C.; Lu, C.; Oh, J. Y.; Wang, Y.; Kim, Y.-H.; Yan, H.; Sinclair, R.; Zhou, D.; Xue, G.; Murmann, B.; Linder, C.; Cai, W.; Tok, J. B.-H.; Chung, J. W.; Bao, Z. Highly stretchable polymer semiconductor films through the nanoconfinement effect. *Science* **2017**, *355* (6320), 59–64. https://doi.org/10.1126/science.aah4496.

[120] Xu, J.; Wu, H.-C.; Zhu, C.; Ehrlich, A.; Shaw, L.; Nikolka, M.; Wang, S.; Molina-Lopez, F.; Gu, X.; Luo, S.; Zhou, D.; Kim, Y.-H.; Wang, G.-J. N.; Gu, K.; Feig, V. R.; Chen, S.; Kim, Y.; Katsumata, T.; Zheng, Y.-Q.; Yan, H.; Chung, J. W.; Lopez, J.; Murmann, B.; Bao, Z. Multi-scale ordering in highly stretchable polymer semiconducting films. *Nat. Mater.* **2019**, *18* (6), 594–601. https://doi.org/10.1038/s41563-019-0340-5.

[121] Nikzad, S.; Wu, H.-C.; Kim, J.; Mahoney, C. M.; Matthews, J. R.; Niu, W.; Li, Y.; Wang, H.; Chen, W.-C.; Toney, M. F.; He, M.; Bao, Z. Inducing molecular aggregation of polymer semiconductors in a secondary insulating polymer matrix to enhance charge transport. *Chem. Mater.* **2020**, *32* (2), 897–905. https://doi.org/10.1021/acs.chemmater.9b05228.

[122] Yin, H.; Zhu, Y.; Youssef, K.; Yu, Z.; Pei, Q. Structures and materials in stretchable electroluminescent devices. *Adv. Mater.* **2022**, *34* (22), 2106184. https://doi.org/10.1002/adma.202106184.

[123] Schlingman, K.; Chen, Y.; Carmichael, R. S.; Carmichael, T. B. 25 years of light-emitting electrochemical cells: a flexible and stretchable perspective. *Adv. Mater.* **2021**, *33* (21), 2006863. https://doi.org/10.1002/adma.202006863.

[124] Filiatrault, H. L.; Porteous, G. C.; Carmichael, R. S.; Davidson, G. J. E.; Carmichael, T. B. Stretchable light-emitting electrochemical cells using an elastomeric emissive material. *Adv. Mater.* **2012**, *24* (20), 2673–2678. https://doi.org/10.1002/adma.201200448.

[125] Wang, J.; Yan, C.; Cai, G.; Cui, M.; Lee-Sie Eh, A.; See Lee, P. Extremely stretchable electroluminescent devices with ionic conductors. *Adv. Mater.* **2016**, *28* (22), 4490–4496. https://doi.org/10.1002/adma.201504187.

[126] Liang, J.; Li, L.; Niu, X.; Yu, Z.; Pei, Q. Elastomeric polymer light-emitting devices and displays. *Nat. Photonics* **2013**, *7* (10), 817–824. https://doi.org/10.1038/nphoton.2013.242.

[127] Gao, H.; Chen, S.; Liang, J.; Pei, Q. Elastomeric light emitting polymer enhanced by interpenetrating networks. *ACS Appl. Mater. Interfaces* **2016**, *8* (47), 32504–32511. https://doi.org/10.1021/acsami.6b10447.

[128] Fratini, S.; Nikolka, M.; Salleo, A.; Schweicher, G.; Sirringhaus, H. Charge transport in high-mobility conjugated polymers and molecular semiconductors. *Nat. Mater.* **2020**, *19* (5), 491–502. https://doi.org/10.1038/s41563-020-0647-2.

[129] Mun, J.; Ochiai, Y.; Wang, W.; Zheng, Y.; Zheng, Y.-Q.; Wu, H.-C.; Matsuhisa, N.; Higashihara, T.; Tok, J. B.-H.; Yun, Y.; Bao, Z. A design strategy for high mobility stretchable polymer semiconductors. *Nat. Commun.* **2021**, *12* (1), 3572. https://doi.org/10.1038/s41467-021-23798-2.

[130] Lu, C.; Lee, W.; Gu, X.; Xu, J.; Chou, H.; Yan, H.; Chiu, Y.; He, M.; Matthews, J. R.; Niu, W.; Tok, J. B.-H.; Toney, M. F.; Chen, W.; Bao, Z. Effects of molecular structure and packing order on the stretchability of semicrystalline conjugated poly(tetrathienoacene-diketopyrrolopyrrole) polymers. *Adv. Electron. Mater.* **2017**, *3* (2), 1600311. https://doi.org/10.1002/aelm.201600311.

[131] Zhang, S.; Ocheje, M. U.; Huang, L.; Galuska, L.; Cao, Z.; Luo, S.; Cheng, Y.; Ehlenberg, D.; Goodman, R. B.; Zhou, D.; Liu, Y.; Chiu, Y.; Azoulay, J. D.; Rondeau-Gagné, S.; Gu, X. The critical role of electron-donating thiophene groups on the mechanical and thermal properties of donor–acceptor semiconducting polymers. *Adv. Electron. Mater.* **2019**, *5* (5), 1800899. https://doi.org/10.1002/aelm.201800899.

[132] Oh, J. Y.; Rondeau-Gagné, S.; Chiu, Y.-C.; Chortos, A.; Lissel, F.; Wang, G.-J. N.; Schroeder, B. C.; Kurosawa, T.; Lopez, J.; Katsumata, T.; Xu, J.; Zhu, C.; Gu, X.; Bae, W.-G.; Kim, Y.; Jin, L.; Chung, J. W.; Tok, J. B.-H.; Bao, Z. Intrinsically stretchable and healable semiconducting polymer for organic transistors. *Nature* **2016**, *539* (7629), 411–415. https://doi.org/10.1038/nature20102.

[133] Wu, H.; Lissel, F.; Wang, G. N.; Koshy, D. M.; Nikzad, S.; Yan, H.; Xu, J.; Luo, S.; Matsuhisa, N.; Cheng, Y.; Wang, F.; Ji, B.; Li, D.; Chen, W.; Xue, G.; Bao, Z. Metal–ligand based mechanophores enhance both mechanical robustness and electronic performance of polymer semiconductors. *Adv. Funct. Mater.* **2021**, *31* (11), 2009201. https://doi.org/10.1002/adfm.202009201.

[134] Ocheje, M. U.; Charron, B. P.; Nyayachavadi, A.; Rondeau-Gagné, S. Stretchable electronics: recent progress in the preparation of stretchable and self-healing semiconducting conjugated polymers. *Flex. Print. Electron.* **2017**, *2* (4), 043002. https://doi.org/10.1088/2058-8585/aa9c9b.

Audithya Nyayachavadi and Simon Rondeau-Gagné

Cross-linking strategies for π-conjugated polymers in organic electronics

Abstract: Organic π-conjugated polymers constitute an important class of semiconducting materials with synthetically tunable optoelectronic and thermomechanical properties. These materials represent an ideal platform for designing emerging electronic and sensing technologies because these materials are also manufacturable at large scale through inexpensive and scalable solution- deposition techniques, thus allowing for the fabrication of new functional technologies at low cost. However, despite these promising features, organic π-conjugated polymer-based technologies suffer from limited lifetime stabilities, and the materials intrinsic solubility in common organic solvents makes advanced fabrication techniques, such as multilayer deposition and nanoscale lithography, challenging. This chapter introduces and reviews emerging strategies that have been utilized by materials scientists for developing cross-linked semiconducting polymers to improve the electronic performance and stability of materials for plastic and soft electronic applications. Throughout, emphasis will be given to designs that utilize supramolecular interactions and chemical motifs incorporated within the side chains and backbone of conjugated polymers that undergo covalent-bond formation, while describing the roles of these interactions and their implications for electronic, processing, and mechanical properties. These highlighted strategies for processing and improving the performance of organic π-conjugated polymers through crosslinking will lead to the development of new materials used in the fabrication of long-lasting and eco-friendly next generation technologies.

1 Introduction to polymeric crosslinking

In the pursuit of modulating the mechanical, thermal, processability, and overall stability of polymeric materials, a large body of research has been conducted over the past centuries in the processing of polymers to improve their utility for various industrial and biomedical applications [1–3]. Among the various techniques developed to control and modulate polymer materials key properties, crosslinking is one of the most used and well-known techniques. Defined broadly as the formation of intermolecular covalent or supramolecular bonds between adjacent polymer chains, crosslinking has become a reliable and highly versatile method for improving key material properties such as thermal stability, surface adhesion, Young's moduli, solid-state morphology and solubil-

Audithya Nyayachavadi, Simon Rondeau-Gagné, Department of Chemistry and Biochemistry, University of Windsor, Essex Centre of Research (CORe), 401 Sunset Ave., Windsor, Ontario N9B 3P4, Canada

https://doi.org/10.1515/9783110757286-002

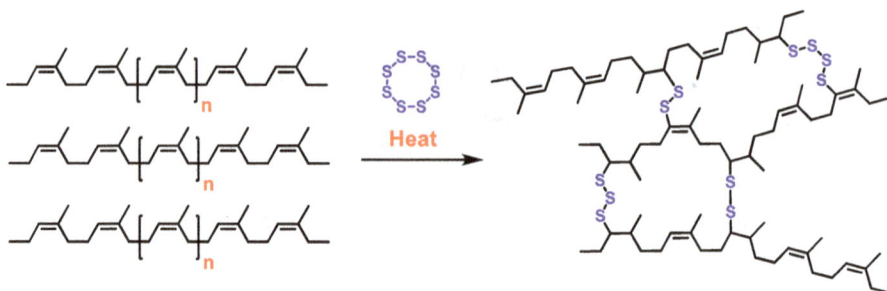

Figure 1: Vulcanization of *cis*-polyisoprene.

ity, molecular weight, dispersity, regioregularity, mechanical toughness, among others [4–10]. One prototypical example of polymeric crosslinking is the vulcanization of rubber (Figure 1), chemically known as polyisoprene, through the addition of sulfur and subsequent heating, which was discovered by Charles Goodyear in the late 1830s [11]. As the urban legend goes, Goodyear was looking for a methodology that would prevent rubber tires from softening and potentially melting on hot roads.

By chance, Goodyear discovered that, by adding S_8 to polyisoprene and heating the mixture, the resulting material became more rigid and resisted melting at higher temperatures. Furthermore, by varying the amount of S_8 added, as well as the heating time, the elastic properties and flexibility of the material could be altered significantly, allowing for unique material properties and applications based on the cross-linking density. After several years of further experimental refinement, vulcanized rubber was being produced for a range of commercial applications including rubber hosing, shoe soles, and car tires.

Built upon this revolutionizing discovery, there is now an extensive library of commercially available cross-linking reagents for diverse polymers with broad applications in industrial, medical, and recreational sectors (Figure 2). The development of cross-linking not only contributed to the rise of polymeric materials in these sectors, but also made it possible to develop many new materials technology, including gels, encapsulants, coatings, adhesives, ultra-stable paints, and nanoparticles, among others [12–15].

The material implications of cross-linking for polymeric materials are numerous. For example, the increase in molecular weight that arises from the binding of adjacent chains leads to a reduction in solubility as materials become more rigid, as well as their altering thermomechanical characteristics such as the glass transition temperature (T_g) and melting temperature (T_m) [16, 17]. Furthermore, as cross-linking results in the formation of 3D network materials, changes in morphology and free volume occur, allowing for highly porous structures capable of "swelling", i. e., absorbing mediums such as liquids or gases for storage, capture, and/or release applications [18, 19].

Over the past century, there has been a concentrated effort to model and measure cross-linking and predict its influence on bulk material properties. Among these, the parameter of cross-linking density or degree of cross-linking refers to the number of

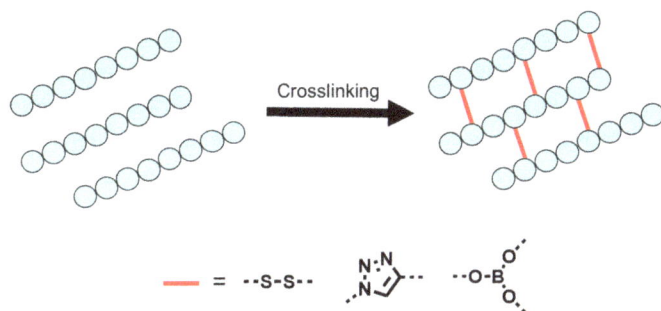

Figure 2: General representation of polymer cross-linking with examples of commonly used crosslinking motifs.

crosslinked bonds in the polymer that is being cross-linked [20]. Often, this parameter is critical for the thermomechanical properties of materials since lower crosslinking densities result in materials behaving as elastomers, while higher densities result in rigid thermosets, solid materials that lack thermal phase transitions [21]. There are several studies and reviews that discuss mathematical modelling of cross-linking density, as well other associated properties such as solvent swelling capacity and gelation point, as well as more specialized concepts such as the role of cross-linking chemical structure and the influence of co-additives [22–26]. Experimentally, there are several techniques that can be used to measure the cross-linking density of polymers, with these measurements usually calculating changes in the physical or thermomechanical properties of materials to look for changes in attributes such as solubility, T_g, and molecular weight [27]. Standard techniques include dynamic mechanical analysis (DMA), rheology, gelation equilibrium swelling, compression tests, Brunauer–Emmett–Teller (BET) measurements, among others [28–33]. Additionally, tests that directly look for changes in chemical structure, such as nuclear magnetic resonance (NMR) and X-ray photoelectron spectroscopy (XPS), have also increasingly become of interest [34, 35]. Given the focus of this chapter on semiconducting materials and related technologies, the reader is strongly encouraged to consult a specialized literature review to learn more about polymer cross-linking as a broad approach in polymer chemistry [36, 37].

2 Introduction to cross-linking of conjugated polymers

2.1 Background

Over the first quarter of the 21st century, a strong demand has emerged for technologies that can be seamlessly integrated into existing societal structures; from buildings to vehicles to the human body, itself, the canvas for platforms utilizing electronics for data acquisition, diagnostics, and ultimately improving efficiency is ever expanding [38]. Collectively dubbed the Internet of Things (IoT), all sorts of household, consumer, and industrial objects are constantly acquiring and transmitting increasingly large amounts of information for analysis and optimization. Current mass-produced electronics are based on silicon, which through a century of innovation has become the *de facto* material for designing leading-edge, high-performance devices [39]. This has enalbed remarkable developments to improve our quality of life, with lasting impacts on education, healthcare, safety, and overall efficiency. Despite these noted achievements, the continued use of silicon as a platform has not been without caveats, notably failing to address the ever-looming associated issues, such as rising energy and material costs, environmental sustainability, and the geopolitical instability related to resource acquisition and expenditures [40–43]. Additionally, silicon has intrinsically poor mechanical conformability and robustness, resulting in technologies often poorly suited for seamless integration into object/surfaces with complex forms [44]. To address these limitations, there has been a rapidly growing interest in developing electronic technologies that forgo the use of silicon as the electroactive material for electronic components in favor of more sustainable, inexpensive, and more customizable materials [45].

Among the various candidates being considered for next generation technologies, organic π-conjugated polymers have proved a promising alternative to silicon for designing so-called "plastic electronics" (technologies whose primary active components are carbon-based) [46]. The employment of conjugated polymers as electroactive materials started in 1977 when Shirakawa et al. doped trans-polyacetylene with halogen vapors, thus successfully developing highly conductive plastic films [47]. Upon the publishing of these results, a new world opened for researchers to achieve the dream of designing electronic materials that transcend the limitations of silicon-based materials, both in terms of economic potential and technological innovation. These efforts have realized the creation of new technologies including organic thin-film transistors, organic photovoltaics (OPVs), and organic light emitting diodes (OLEDs) [48]. In the near half century since, countless chemical innovations have been and continue to be reported towards designing high-performance technologies, notable examples of which include poly(*p*-phenylenevinylene), polyaniline, and polypyrrole [49–51]. Compared to traditional silicon electronics, semiconducting polymers (SPs) confer some distinct advantages, namely: their inherent solution processability through the incorporation of solubilizing alkyl side chains as well as an extended conjugated backbone, allowing for improved mechanical properties, primarily due to chain entanglements; greater inter- and intramolecular interactions; and semicrystalline nature when cast as thin films [52, 53]. Consequently, a large body of research has been dedicated to understanding the structure–property relationships of SPs for improving both their mechanistic and optoelectronic qualities, with designs successfully achieving charge-carrier mobilities >10 cm^2 V^{-1} s^{-1}, electronic metrics that are within the realm of smaller molecular counterparts and of amorphous silicon [54].

While breakthroughs continue to be achieved in terms of improving the electronic performance of SPs, the integration of π-conjugated polymers into consumer, biomedical, and industrial technologies requires greater fundamental understanding, both from a chemical and engineering perspective, to increase their effectiveness in electronic performance, mechanical compliance, and long-term stability [55, 56]. Typically, π-conjugated SPs exhibit low long-term stability and poor surface adhesion due to their soft and semicrystalline nature, resulting in changes in solid-state morphology and nano/microstructures over time, resulting in decreased performance and device failure [57]. Additionally, for applications that require constant mechanical stress and movement (including bioelectronics), these materials require additional rigidification to prevent delamination and consequent failure.

Among the various strategies for improving the properties of conjugated polymers, cross-linking is an emerging approach for offering expanded dimensions of control in enhancing the mechanical compliance, electronic performance, solubility, and long-term stability of materials from chemical-structure and morphological perspectives [58]. In general, the concept of cross-linking has become part and parcel of the field of polymer chemistry and has been shown to impart novel properties to conventional plastics and rubbers, resulting in materials that become stretchable, flexible, porous, and resistant to environmental stress, among other attributes. A similar underlying principle is applied for conjugated polymers that are used for conventional polymers, namely, cross-linking to initiate the formation of new bonds (covalent or noncovalent) between adjacent, separate polymer chains, with said bonds being perpendicular to that of the polymer backbone [59] (Figure 3). By initiating the formation of new bonds across multiple polymer chains, dramatic effects on physical properties occur because now the chains are "frozen" or prevented from moving freely, ultimately changing material rigidity, melting point, solubility, among other characteristics, just as they would in conventional polymers [60].

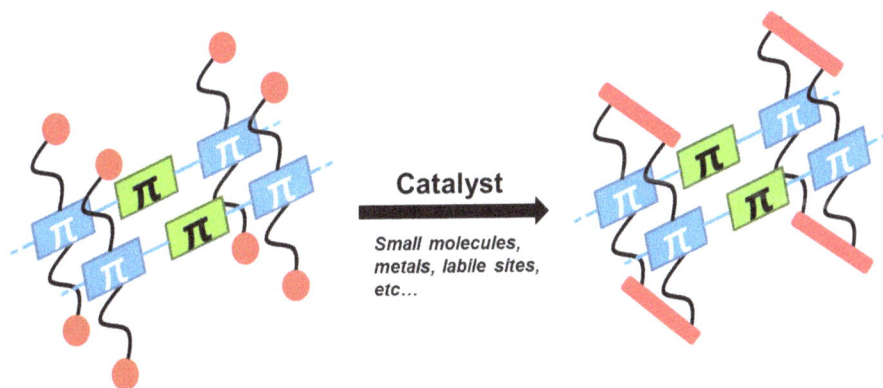

Figure 3: Cross-linking of π-conjugated polymer through the lateral side chains.

This chapter discusses several strategies for the design and synthesis of π-conjugated polymers for organic electronics that can undergo cross-linking, while also surveying the influence of crosslinking on the processing, mechanical, and electronic properties of the resulting materials. It is important to note that, concurrently, while there are design strategies for crosslinking conjugated polymers for drug delivery, diagnostic, and photocatalytic applications, the scope of this review will focus on cross-linked materials designed to targeted organic electronic applications [61, 62]. Therefore, only the fabrication and conception of cross-linking approaches to conjugated polymers and their incorporation in organic electronics will be discussed.

3 Noncovalent cross-linking of conjugated polymers

Within the realm of materials chemistry, the utilization of supramolecular interactions in small molecules, composites, and polymers to modulate their thermomechanical and electronic properties has become an efficient and popular design strategy because of the demonstration that careful control of these interactions has proved to strongly influence material stability, while also enhancing targeted properties such as conductivity and self-healing characteristics [63]. Thus, it is of no surprise that this ubiquitous approach has been applied to π-conjugated polymer systems as wel because it has been demonstrated that incorporation of dynamic noncovalent cross-linking interactions between polymer chains can improve the mechanical and self-healing capabilities of materials, while also influencing their solid-state morphology through the formation of discrete nanostructures, thus also influencing their anisotropic and electronic properties [64].

3.1 Hydrogen bonding-induced cross-linking

A hydrogen bond is a weak electrostatic attraction between a hydrogen atom bonded to an electronegative atom and another nearby electronegative atom (oxygen, fluorine, nitrogen), which leads to the formation of a partial positive and partial negative charge interaction. Falling under the class of dipole–dipole interactions, the range of energy it takes to dissipate a hydrogen bond varies from 1 to 40 kcal/mol, making them stronger than most van der Waals interactions but significantly weaker than that of a covalent bond [65]. These attributes provide unique implications for materials since hydrogen bonds can reversibly reform and reorganize themselves at elevated temperatures due to their strong dipole interaction [66]. The implications of hydrogen bond-driven cross-linking on materials are numerous since they can drive material structural self-assembly, enhance mechanical properties, impart self-healing character, and alter solubility and processing conditions [67]. Furthermore, it has been shown that hydrogen bonding alters the optoelectronic properties of conjugated molecules and

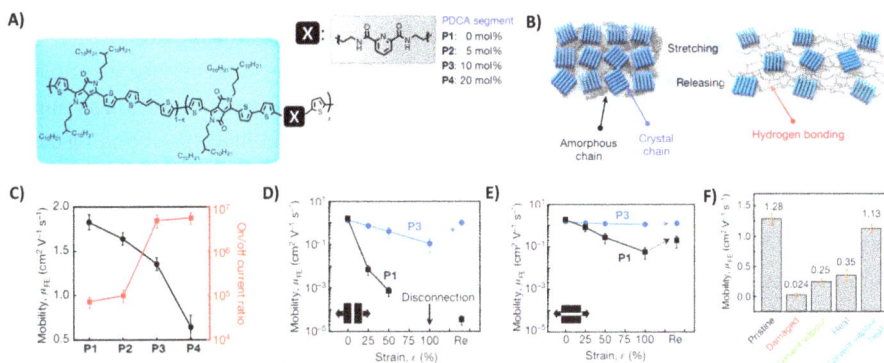

Figure 4: A) Chemical structures of diketopyrrolopyrrole polymers **P1** to **P4** containing various mol ratios of PDCA hydrogen-bonding segments. B) Representative mechanism for enhancement of stretchability in conjugated polymers via hydrogen bonding. C) OFET mobility μ_{FE} and on/off current ratios of **P1** to **P4**. D) Change in mobility μ_{FE} of **P1** and **P3** as a factor of strain in parallel and E) perpendicular directions across charge transport. F) Change in mobility of **P3** upon various damage and healing treatments. Adapted with permission from reference [72]. Copyright 2016 Nature Publishing Group.

polymers, offering the additional dimension of tunability when designing materials. For these reasons, the incorporation of hydrogen-bonding motifs in conjugated polymers is a widely employed and continued approach for designing high- performance plastic electronics with targeted applications. There are many review articles that discuss the roles of hydrogen bonding and its implications in conjugated polymers in detail [68–71]. For these reasons, we will highlight some key studies of hydrogen bonding as a cross-linking methodology and some of the various attributes imparted.

One of the first examples of incorporating hydrogen bonding into conjugated polymers was reported in 2016 by Oh et al. [72]. In this investigation, a series of conjugated diketopyrrolopyrrole (DPP) polymers that were regiorandomly incorporated with different ratios of 2,6-pyridine dicarboxamide (PDCA) conjugation breaking segments (Figure 4) was synthesized. The PDCA chemical motif was selected to facilitate intrachain hydrogen bonds to develop stretchable and self-healing conjugated polymer transistors, as represented in Figure 4B.

It was observed that, by increasing the content of PDCA within the backbone, a relative decrease in charge-carrier mobility occurred (Figure 4C) as well as a decrease in Young's modulus, revealing a plasticizing effect particularly important for mechanical properties. When evaluating **P3** (which contained 20 % PDCA content) against **P1** (which contained no PDCA moiety), as shown in Figures 4D and 4E, it was shown that **P3** possessed greater retention of charge mobility across 100 % strain in both the parallel and perpendicular charge-transport directions, exhibiting minimal loss in comparison to **P1** which lacked hydrogen bonding. These results demonstrated that the inclusion of hydrogen bonding helped preserve the electronic integrity of the polymer under greater mechanical strain. Furthermore, it was shown that the incorporation of PDCA imparted a self-healing character to the polymer thin films, which upon cutting and subsequent annealing under solvent vapor conditions demonstrated that the polymer was able to maintain 88 % of its original charge mobility in comparison to the pristine film as shown in Figure 4F. This observation showed that the incorporation of hydrogen bonding could also serve to increase the lifetime of semiconductive materials as well, which when combined with enhanced mechanical properties provides a potential avenue for technologies that are exposed to constant movement, such as skin-inspired electronics. Following from this study, the attempts to study hydrogen bonding as a platform for self-healing and mechanically compliant electronics has grown, with various motifs being investigated within the side chain and backbone of conjugated polymers, including amides, ureas, nucleobases, and alcohols, with the aim of understanding their role in conjugated polymer design [73–76].

The role of hydrogen bonding as a form of supramolecular cross-linking has been shown to also influence other parameters important to organic electronics, namely, control over solubility and resistance to chemical etching for multilayer deposition and lithographic applications [77]. The predominant methodology of this approach is the incorporation of labile protecting groups that act as solubilizing motifs within the conjugated polymer, which upon casting into thin films are capable of being cleaved chemically or thermally, exposing the hydrogen bonding motif that will interact with other exposed adjacent polymer chains, thus supramolecularly cross-linking the polymers and promoting aggregation, and providing greater resistance to chemical etching [78]. A recent example of this approach was reported by Hao Guo et al. who announced the synthesis of an isoindigo (iI)-based conjugated polymer that incorporated various ratios of the thermally cleavable *tert*-butyloxycarbonyl (Boc) protecting group as the side chains, as shown in Figure 5A [79].

Upon confirmation of the removal of the Boc groups through FTIR (Fourier transform infrared spectroscopy) and TGA (thermogravimetric analysis), multilayer deposition of the polymer films was demonstrated by repeated casting of thin films onto a

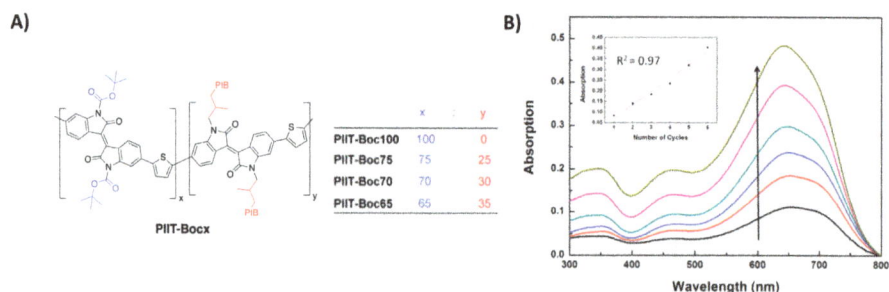

Figure 5: A) Chemical structure of iI-containing polymers with various ratios of Boc cleavable side chains. B) UV-Vis spectra of **PIIT-Boc70** film by "casting–annealing–casting–annealing" cycles (inset graph: the relationship between the number of cycles and absorption at 600 nm). Adapted with permission from reference [79]. Copyright 2016 Royal Society of Chemistry.

glass slide and subsequent annealing to immobilize the films onto the surface upon removal of the Boc groups, thus conferring chemical resistance, which was quantified by a linear relationship in increasing absorbance via UV-Vis spectroscopy as shown in Figure 5B. Furthermore, upon thermal cleavage, it was demonstrated that removal of Boc protecting groups showed negligible change in thin-film transistor mobility, successfully demonstrating this approach as a methodology for developing multilayer optoelectronics. Notably, Ji et al. reported a similar approach for Boc chain removal by using a DPP-based polymer system and observed an improvement in charge-carrier mobility from $0.56\,\mathrm{cm^2\,V^{-1}\,s^{-1}}$ to $0.91\,\mathrm{cm^2\,V^{-1}\,s^{-1}}$, indicating that this type of cross-linking can be used not only to tune solubility but also to enhance electronic performance through careful selection of chemical design [80]. For these reasons, the Boc cleavage strategy has been widely employed in many different conjugated polymer systems for device applications, including OFETs and OPVs. Apart from the amine protecting Boc group, other cleavable groups to expose different hydrogen bonding motifs capable of supramolecular cross-linking have been utilized as well, such as acetals, ethers, and esters [60]. Careful selection of the protecting group can confer unique advantages based on cleavage parameters. Particularly, Smith et al. reported a polythiophene derivative with phenol side chains in the 3 position that were protected with o-nitrobenzyl ethers capable of being photocleaved to form organic solvent-resistant films [81]. Due to the photocleavage mechanism, these materials were able to be photopatterned *via* the use of a negative photoresist and still maintained thin-film transistor performance comparable to the uncross-linked materials. Thus, the use of hydrogen-bonding crosslinking as a methodology for controlling the solubility and chemical resistance of materials is a versatile approach that continues to be explored.

Another advantage of hydrogen bond-driven cross-linking is the ability to improve the electronic performance of conjugated polymers. One defining example of this was reported by Yao et al. who synthesized a series of DPP-based polymers that possessed varying contents of hydrogen-bonding urea side chains (as shown in Figure 6A), with the hypothesis being that the urea would provide a directionality in hydrogen bonding capable of cross-linking adjacent polymer chains together to promote crystallinity, critical for good electronic charge transport in devices, as represented in Figure 6B [82].

Upon synthesis, the urea-containing polymers were cast into thin films for bottom-gate bottom-contact OFETs. Once annealed, it was found that **PDPP4T-3**, which contains 10 % urea side chain achieved, a hole mobility (μ_h) of $13.1\,\mathrm{cm^2\,V^{-1}\,s^{-1}}$, which is, at this time, one of the highest mobilities reported for DPP-based polymers. In comparison, the polymer without any hydrogen bonding, **PDPP4T**, achieved a hole mobility of $3.4\,\mathrm{cm^2\,V^{-1}\,s^{-1}}$, significantly less than the hydrogen bonding-containing polymer. This observation was also consistent with the 20 % and 30 % urea containing polymers, which were also observed to have higher charge mobilities under similar processing conditions in comparison to the reference polymer. These results indicated that the conjugated polymers could have enhanced charge-transport characteristics due to the presence of hydrogen bonding. These results were further elaborated by looking at the

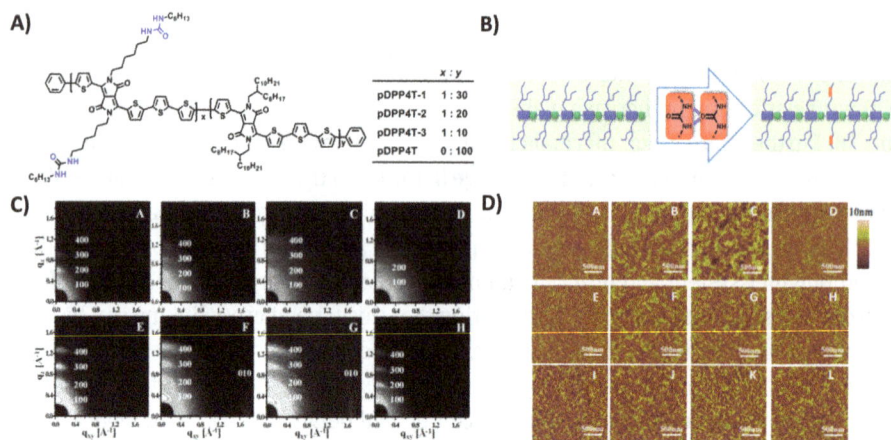

Figure 6: A) Chemical structures of **PDPP4T-1, PDPP4T2, PDPP4T-3**, and **PDPP4T** B) Illustration of the design rationale for incorporation of urea groups C) Grazing incidence X-ray diffraction (GIXRD) patterns of **PDPP4T-1** (A, E), **PDPP4T-2** (B, F), **PDPP4T-3** (C, G), and **PDPP4T** (D, H) deposited on OTS-modified SiO$_2$/Si substrates at room temperature (up) and after thermal annealing at 100 °C (bottom). D) AFM height images of thin films of **pDPP4T-1** (A, E, I), **PDPP4T-2** (B, F, J), **PDPP4T-3** (C, G, K), and **PDPP4T** (D, H, L) deposited on OTS-modified SiO$_2$/Si substrates at room temperature (up) and after thermal annealing at 100 °C (middle) and 120 °C (down). Adapted with permission from reference [82]. Copyright American Chemical Society, 2016.

morphology of the materials using GIXRD and AFM, as per Figures 6C and 6D respectively. Notably, it was observed that, upon annealing at 100 °C, the hydrogen bonding-containing polymers were significantly more crystalline than the reference polymer, as observed by greater intensity higher-order diffraction peaks and the appearance of the 010 peak for the 10 % and 20 % containing urea polymers, which is indicative of ordered pi-stacking interactions. Complimentary to these results, AFM was also conducted, and it was observed that the urea containing polymers showed a greater fibrous morphology upon spin coating and annealing into thin films, indicative that the presence of hydrogen bonding-interactions promotes the self-assembly of conjugated polymers that are critical for good electronic and mechanical properties. Furthermore, the polymers were also tested in OPVs as electron donors after blending with PC$_{71}$BM as the electron acceptor, and it was observed that the hydrogen bonding-containing polymers had greater power conversion efficiencies compared to the reference polymer, indicating that the fibrous morphology afforded by hydrogen bonding enhanced the charge transport characteristics of the materials and yielded better performing devices. Overall, this study demonstrated the potential of hydrogen bonding as a tool for controlling the morphology and enhancing the electronic properties of materials.

Additional studies have been reported with other hydrogen-bonding motifs, such as thymines and carbohydrates, that have been shown to improve the electronic properties of conjugated polymer systems as well [83, 84]. While hydrogen bonding as a cross-linking approach remains popular for enhancing the various properties of organic semiconductors, there are still various motifs that also have yet to be explored for understanding and optimizing the self-assembly implications of such motifs for targeted device application.

3.2 Host–guest interaction for cross-linking

Another studied approach for supramolecular cross-linking of conjugated polymers is the exploitation of host–guest interactions. The principle behind this approach is that a conjugated polymer can be chemically designed to contain either a host or guest motif, upon which a complimentarily designed additive will be introduced to supramolecularly bind the adjacent polymer chains together to build a 3D cross-linked network [85]. One prominent example in the literature reported by Pappalardo et al. is the synthesis of a poly(p-phenyleneethynylene) containing terminal calix[5]arene host side chains that are capable of being supramolecularly cross-linked using 1,10-decanediyldiammonium guest linkers [86]. In the study, it was shown that, upon variation of concentration of the 1,10-decanediyldiammonium linker, the type of polymer self-assembled morphology that was observed could change, with smaller equivalents of linker favoring cross-linked networks, while larger equivalents favor "capping" of the poly(p-phenyleneethynylene) as single chains, which was confirmed through proton NMR experiments and AFM. Interestingly, it was shown that, by varying the concentration of linker, the fluorescence intensity of the resulting material could also be manipulated, with fluorescence quenching occurring after adding greater than 0.5 eq of the linker. Finally, it was shown that through titration of the acid/base, the host–guest interaction was reversible, overall confirming this methodology as a dynamic approach to control the optoelectronic and structural properties of conjugated materials. Subsequently, this approach has also been explored for other host–guest cross-linking interactions, including dibenzo[24]crown-8 and pillar[5]arene towards improving the electronic properties of materials and providing novel applications such as sensing [87, 88].

Notably, Fu et al. reported a unique approach for cross-linking via host–guess interactions by incorporating a guest motif directly within the backbone of a conjugated polymer through the synthesis of polycarbazoles capable of being cross-linked by acting as a guest to a bis[alkynylplatinum(II)]terpyridine based linker [89] (Figure 7). Upon complexation, the crossl-inked polymer was shown to have decreased fluorescence emission, which was attributed to an energy-transfer process between the donor carbazole and acceptor bis[alkynylplatinum(II)]terpyridine linker, showing that the cross-linking interaction could alter the optoelectronic properties of the material. Furthermore, the addition of the bis[alkynylplatinum(II)]terpyridine linker was observed to facilitate the formation of thermally reversible organogels in chloroform solvent, whose viscosity could be controlled through varying the concentration of linker. This observation not

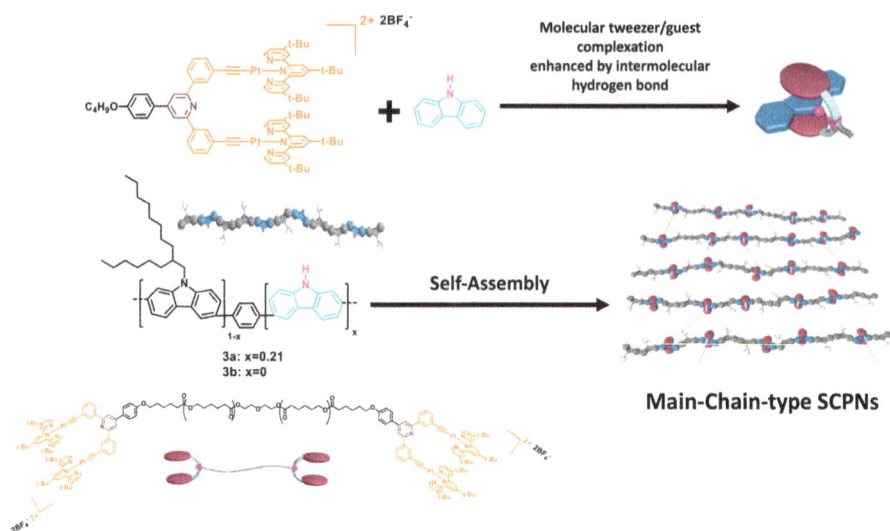

Figure 7: Schematic representation for the formation of main-chain-type supramolecular π-conjugated polymer networks between polycarbazoles and bis[alkynylplatinum(II)]terpyridine linker. Adapted with permission from reference [89]. Copyright American Chemical Society, 2017.

only shows great promise for controlling the mechanical properties of conjugated polymers but also for manipulating solution state properties as a potential for ink formulations in printed electronics.

Host–guest interactions as a cross-linking strategy remains a fascinating approach for controlling the electronic, thermomechanical, and processing capabilities of conjugated polymers. Thus, it is to be expected that future work will involve the incorporation of other types of interactions as well understanding their implications in organic electronics. Furthermore, the possibility of designing novel linkers that can incorporate unique motifs within conjugated polymer systems upon cross-linking shows great potential for designing multifunctional electronic technologies.

3.3 Metal–ligand interactions for cross-linking

In a similar vein to host–guest style crosslinking, the application of metal–ligand (M–L) interactions as an approach to supramolecular crosslinking of conjugated polymers is a promising strategy for controlling the mechanical and electronic properties of materials for organic electronic application. The principle behind this strategy is based on the incorporation of a ligand motif within the polymer chemical structure, so that the polymer chains can be crosslinked through the addition of a metal source of interest [90]. This approach has a lot of implications for material processing and performance since the incorporation of different metal centers and their affinity for different ligands, as well as their position in the polymer structure, has the potential to drastically change the chemical, morphological, mechanical, and electronic properties of a system [91].

One of the earlier examples of M–L cross-inking in conjugated polymers was by Welterlich et al., who reported the synthesis of a poly(phenylene-*alt*-fluorene) with 2,6-bis(10-methylbenzimidazolyl) pyridine (bip) ligands capable of divalent metal coordination (Zn(II)and Cu(II)), which was confirmed through UV-Vis and fluorescence spectroscopy [92]. Interestingly, despite the coordination, the cross-linked polymers were found to be soluble and processable in thin fluorescent films. Moreover, the complexation was found to be controllable using specific organic solvents, making the strategy a potential avenue for substrate patterning of thin films. A notable example of M–L cross-linking is the incorporation of pyridine-2,6-dicarboxamide (PDCA) conjugated breaking units within the backbone of a DPP based semiconducting polymer which could coordinate with Fe(II) ions, which would subsequently oxidize into Fe(III) to create cross-inked polymers [93].

As reported by Wu et al., DPP-based polymers were synthesized containing different percentage ratios of the PDCA ligand unit within the backbone of the polymer structure, which were subsequently complexated through the addition of NaH and the $Fe(BF_4)_2$ metal source, which was confirmed through XPS and IR Spectroscopy. Interestingly, it was shown that, even at 30 %, the coordinated materials retained their solubility in common organic solvents, demonstrating that the cross-linking did not impact processing capabilities. The coordinated polymers were subsequently tested in OFETs, where it was observed that the coordination cross-linking improved charge-mobility performance, with mobility nearly doubling with respect to the uncross-linked polymer at a 20 % PDCA ratio, as shown in Figure 8B, which was attributed to the doping effect of the Fe(III) ions affecting the On current of the materials (Figure 8C). Furthermore, it was shown that the coordinated polymers improved the overall device stability of the mate-

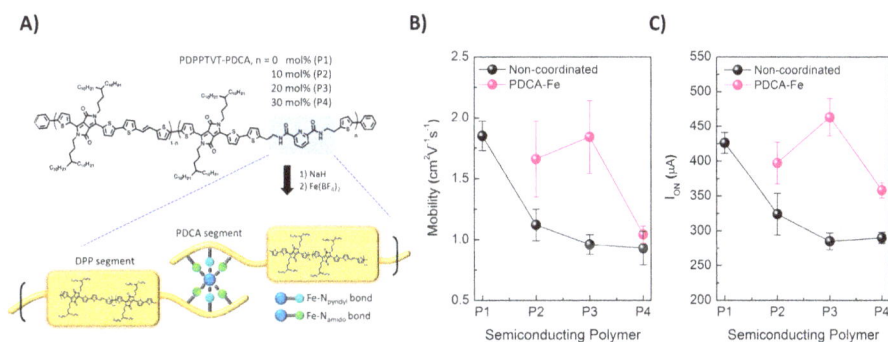

Figure 8: A) Schematic representation for the formation of supramolecular cross-linked π-conjugated DPP-based polymer networks between using pyridine-2,6-dicarboxamide conjugation- breaking spacer units and Fe(II) ions. B) Hole mobility. C) On-current comparisons of non-coordinated and coordinated polymers. Adapted with permission from reference [93]. Copyright Wiley, 2021.

rials through continuous testing of high current on state and low current off state. These findings demonstrate that M–L coordination cannot only improve the performance of organic semiconductors but also increase their overall lifetime potentials.

The implications of metal coordination as a cross-linking methodology remains an emerging and still relatively unexplored approach for application in organic electronics. With some key advantages, namely, the ability for materials to remain solution processable upon crosslinking, as well as the ability to affect electronic properties upon selection of ligands and metal ions, it is expected that future research will seek to understanding the implications of M–L crosslinking on material electronic properties and to begin to investigate their implications on material mechanical properties, such as stretchability and self-healing character.

3.4 Ionic cross-linking

Ionic cross-linking occurs when a polymer with side chains containing a charged functional group interacts with a functional group of the opposite charge [94]. This crosslinking technique has been particularly exploited for biopolymers, including glycogen and chitosan, for drug-delivery and tissue-engineering applications, because careful selection of molecular additive and ion pairs can be used to control properties such as gel viscosity, conductivity, and cross-linking reversibility, all of which play a role when trying to target specific industrial and biomedical applications [95]. Due to the tenability, as well as the ability to alter solid-state morphology, mechanical properties and stability, ionic cross-linking has slowly become an emerging pathway for organic optoelectronics through their incorporation in conjugated polymers. Tian et al. reported the synthesis of a polyphenyl acetylene (PPA) that was chemically modified with a *para*-(4-methylpiperazin-1-yl)methanone motif to act as the "side chain" [96]. Upon dissolving the polymer into solution, suberic acid and 1,8 dibromo octane were added as crosslinkers to facilitate the ionic cross-linking, as evidenced by solvent swelling into the gel state, and further confirmed using XPS and molecular weight data. Notably, it was observed that, in comparison to the 1,8 dibromo octane, the suberic acid-cross-linked network showed greater thermal stability with a higher glass-transition temperature when measured using differential scanning calorimetry. Furthermore, the morphology of the cross-linked polymers was investigated using Brunauer–Emmett–Teller (BET) and Barrett–Joyne–Halenda methods, and it was observed that the suberic acid-cross-linked polymer exhibited an SA_{BET} of 16 m^2/g, nearly three times greater than what was observed with the 1,8 dibromo octane cross-linker. These results indicated that, through careful selection of ionic cross-linking pairs, it is possible to control the morphology and stability of polymeric materials. Furthermore, it was shown that in comparison to the uncross-linked PPA, the suberic acid-cross-linked polymer has a tenfold increase in current density performance, demonstrating that this ionic cross-linking methodology has a positive impact on material electronic properties.

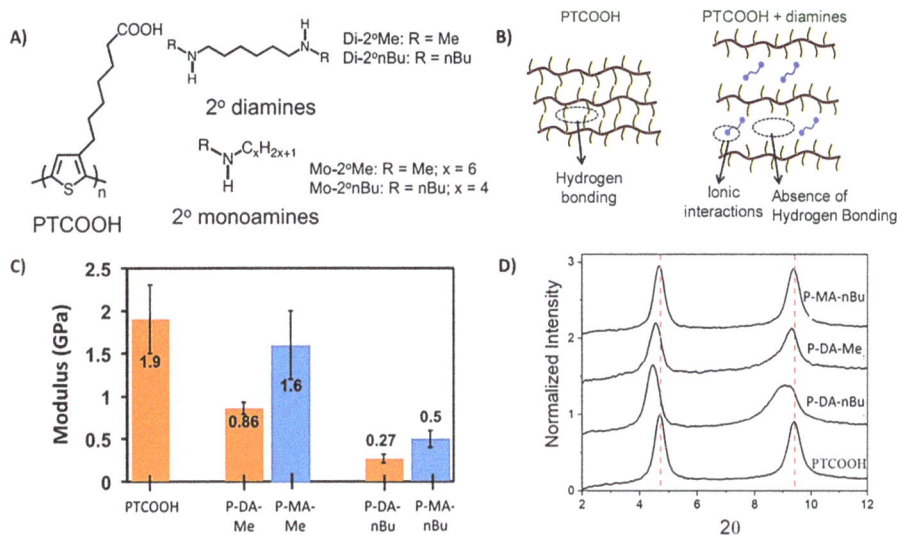

Figure 9: A) Chemical structure of polythiophene with carboxylic acid side chains and amine additives used in the study. B) Representation of diamine additives on supramolecular structure of polymer chains. C) Young's moduli of polymers with amine additives. D) pXRD spectra of polymers and amine additives. Adapted with permission from reference [97]. Copyright Elsevier, 2020.

Notably, Shinde et al. reported an investigative approach to understanding the implications of ionic cross-linking additives on the mechanical properties of semiconductive polymers by designing a 3-heptyl carboxylic acid polythiophene (PTCOOH) that was capable of ionic cross-linking upon the introduction of diamines, as shown in Figure 9A [97].

The hypothesis behind this approach was that, while the carboxylic acid polymer is capable of supramolecular cross-linking through hydrogen bonding, by incorporating diamine groups (with monoamine counter parts acting as controls), the ionic cross-linking could be facilitated and thus change the packing arrangement of the polymers (Figure 9B) and have a direct impact on the mechanical properties of the polythiophenes, namely, the young's modulus. As shown in Figure 9C, it was observed that the incorporation of the diamines drastically reduced the Young's modulus of the PTCOOH from 1.9 GPa to 0.86 GPa with the methyl substituted diamine, and to 0.27 GPa with the n-butyl substituted diamine. These results demonstrated that the incorporation of amine cross-linkers had a plasticizing effect on PTCOOH, and, by varying the structure of the cross-linker such as its bulkiness, it is possible to tune the degree of plasticization. This was further analyzed by powder x-ray diffraction (PXRD) as seen in Figure 9D, which showed that the introduction of the diamines lowered the crystallinity of PTCOOH as shown by the increase in peak broadness at 2θ, as well as a peak shift downward, which indicates an increase in pi-stacking distance between adjacent polymer chains, with the butyl substituted diamine showing the largest pi-stacking distance. Overall, this study demonstrates that ionic cross-linking is a powerful platform for tuning the mechanical characteristics of organic semiconductors.

Due to the recently reported incorporation of ionic cross-linking for conjugated polymers as a methodology, there are many factors and parameters that need to be investigated before incorporation into organic electronic materials on a large scale. Namely, future studies will have to investigate the implications of ionic charges in semiconductor devices such as OFETs, OPVs, and OECTs. However, the benefits of ionic cross-linking are numerous, such as the commercial availability of many cross-linkers and the relatively simple synthetic designs of many ionic pairs, as well as the ability to create biocompatible materials for medical applications. Thus, it is expected that ionic cross-linking will become more prominent in the field of conjugated polymers.

4 Covalent cross-linking of conjugated polymers

Similarly to supramolecular cross-linking strategies, the covalent cross-linking of π-conjugated polymers has also become an attractive post-functionalization approach for modulating the optoelectronic and thermomechanical properties of materials. In contrast to supramolecular interactions, covalent-bond formations through cross-linking are often not thermally or mechanically reversible (initiators and chemical additives notwithstanding), so it is expected that the formation of these more robust bonds result in significantly more rigid networks in comparison to supramolecular cross-linking [98]. Therefore, this approach to cross-linking is not as susceptible to changes in environmental conditions, such as pH, temperature, or humidity, making this strategy ideal for environments where electronic devices undergo constant acute mechanical stress and would require longer lifetimes [98]. Additionally, through careful selection of the cross-linking strategy, it is possible to covalently connect polymer chains with highly desirable motifs that have shown promise in other systems for their mechanical and electronic characteristics, thus ultimately generating hybrid electronic material films that show both useful mechanical and optoelectronic properties. For these reasons, covalent crosslinking has been shown to be a promising avenue for designing long-lifetime plastic electronics with robust mechanical properties. In the following section, we discuss some prominent classes of covalent cross-linking strategies for conjugated polymers and highlight the unique advantages each one confers on final material properties.

4.1 Photo-cross-linking

A subset of covalent cross-linking for conjugated polymers that has received notable attention is the incorporation of photoreactive motifs to the backbone or side chain of conjugated polymers. This methodology is particularly advantageous for lithography and multilayer deposition because the solubility of materials can be readily controlled and localized via thin-film deposition [99]. For these reasons, photo-cross-linking has been widely investigated and optimized in various organic semiconductor systems, with some key strategies highlighted next.

A common strategy is the usage of additives to initiate photo-cross-linking, often through the incorporation of vinyl or acrylate containing side chains with different additives such as thiols, nitriles, alkyl halides, and diazirines [100–102]. Among these approaches, photo-cross-linking *via* incorporation of azide additives has gained attention in recent years. A leading example of this approach reported by Kim et al. was the employment of a 4-armed perfluorophenyl azide crosslinker (often dubbed "4Bx") [103]. Compared to other cross-linking additives, the advantage of 4Bx is its mechanism of insertion to aliphatic C–H bonds upon exposure to UV irradiation, making it to be widely applicable to a wide range of organic materials without need for significant modification of chemical structures. As reported, the high efficiency of the insertion reaction allows for notable alterations in thin film properties of both DPP and naphthalene diimide (NDI) based polymers at low additive loading of 1 wt %, conferring resistance to chemical etching and the ability to be photo-patterned, as well as maintenance of charge-carrier mobility in OFETs. Notably, it was shown that the crosslinking reaction led to increase in the elastic modulus, resulting in more brittle films. The scope of this crosslinker was later expanded upon by Chen et al., where 4Bx was applied to a library of 11 different organic semiconducting polymers to increase the material robustness towards thermal, mechanical, and environmental degradation in OPVs [57].

Notably, it has been shown that, through rational chemical design of perfluorophenyl azide crosslinkers, different mechanical attributes can be imparted onto crosslinked organic semiconductors. This hypothesis was evaluated upon by Wang et al. who reported the effects of a series of bis perfluorophenyl azide crosslinkers (linear crystalline vs. branched amorphous) to understand the effects of each on the mechanical and electronic properties of a DPP-based polymers, as depicted in Figure 10A [104].

As shown in Figure 10B, the fracture strain of the linear crosslinked films was improved twofold, whereas the branch crosslinkers showed a fourfold increase, with the presence of the hydrogen bonding amide groups not displaying any significant impact on the fracture onset. As shown in Figure 10C, the dichroic ratio (R) increased initially for the neat semiconductor and plateaued after $\varepsilon = 25\%$. In comparison, the cross-linked films showed a steady linear increase in R with strains of $\leq 100\%$, with the 5 branch having a maximum R of 2.9. Finally, as depicted in Figure 10D, the branched non-hydrogen bonding cross-linking showed the greatest retention of mobility upon 5,000 cycles of stretching and relaxation at $\varepsilon = 25\%$, demonstrating that careful design of cross-linker structure can impact material stability. Using a similar rationale, Zhang et al. reported the employment of a perfluorophenyl azide end-capped polybutadiene cross-linker for DPP and indacenodithiophene (IDTBT)-based polymers [105]. By crosslinking IDTBT in

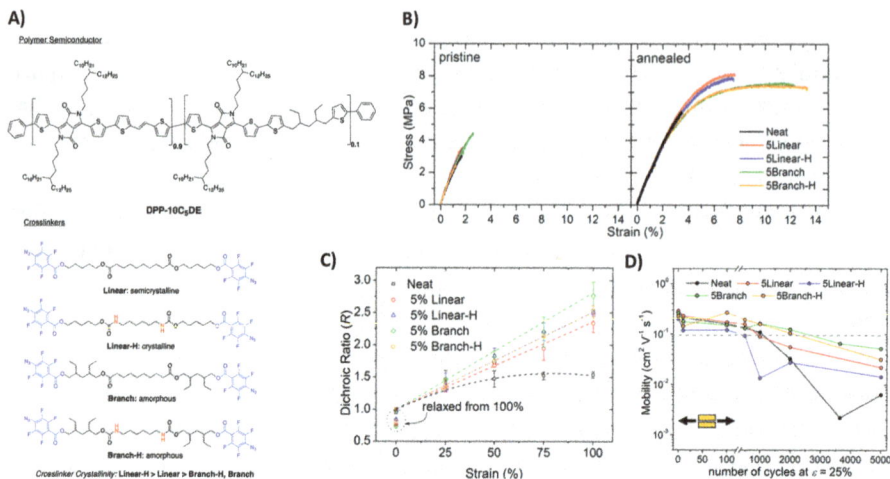

Figure 10: A) Chemical structure of the four cross-linkers and the DPP-based polymer. B) Stress–strain curve of pristine polymer-cross-linker blends (left) and cured polymer films obtained by the film-on-water (FOW) technique. C) Dichroic ratio (R) of cross-linked films under different strains. D) Average mobility of relaxed polymer films after various stretching cycles at ε = 25 % perpendicular to charge transport. Adapted with permission from reference [104]. Copyright American Chemical Society, 2019.

a rubber matrix, Zheng et al. were able to produce ultrathin films with an intrinsic elasticity of 70 %, which could maintain hole mobilities of $1\,\mathrm{cm^2\,V^{-1}\,s^{-1}}$ after 1,000 cycles of stretching to 50 % strain. As an evolution of this approach, Gao et al. reported a cross-linking strategy for a DPP-based polymer that requires no azide additive, by directly incorporating terminal azide groups within the branched solubilizing side chains of their SP [106].

As shown in Figure 11A, by designing a single component material with the azide side chain, referred to as **PDPP4T-N₃**, the polymer was able to be directly photopatterned to form good resolution designs upon exposure to 365-nm light and chemical etching, as shown in Figure 11B. Subsequently, the cross-linked thin films were then tested in bottom-gate–bottom-contact type OFETs and were observed to have an average charge carrier hole mobility of $0.61\pm0.10\,\mathrm{cm^2\,V^{-1}\,s^{-1}}$, which was shown to be lower than the uncross-linked films which had an average mobility of $1.06 \pm 0.32\,\mathrm{cm^2\,V^{-1}\,s^{-1}}$. This discrepancy was attributed to the formation of secondary amino groups within the thin-film network that occur upon cross-linking, which could potentially act as charge traps. Interestingly, by taking advantage of the insolubility that occurs upon cross-linking, the investigators deposited an electron efficient n-type polymer, **F₄BDOPV-2T**, on top of the hole transporting **PDPP4T-N₃** to develop a multilayer device that acts as a complementary inverter as shown in Figures 11C and 11D. The voltage transfer curve was observed to display a distinct switching-action around 20 V with a gain value of 68, indicating the successful formation of a logic-inverter device using two distinct polymers. These re-

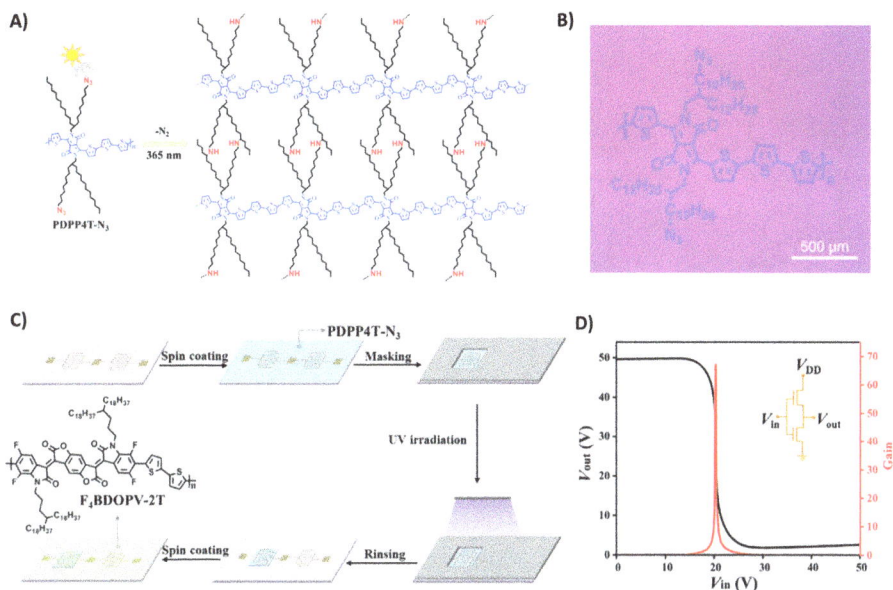

Figure 11: A) Chemical structure of single component semiconducting photoresist **PDPP4T-N₃** illustration of the cross-linking mechanism. B) Optical microscope images of patterns of **PDPP4T-N₃**. C) Illustration of the complementary-like inverter fabrication; the chemical structure of **F₄BDOPV-2T**. D) The corresponding voltage transfer and gain characteristics measured under a nitrogen atmosphere, $V_D = 50\,V$. The inset shows the circuit diagram. Adapted with permission from reference [106]. Copyright Wiley, 2022.

sults demonstrated that, through covalent cross-linking, new device architectures are realizable for the development of multifunctional electronics.

Another notable photo-cross-linking strategy is the dimerization of side chains through the incorporation of motifs that undergo photo-initiated cycloaddition reactions. The hypothesis behind this approach is that the formation of interchain cyclobutanes between adjacent polymer chains can covalently stabilize polymer thin films and blends for better tolerance to chemical etching, while also altering the mechanical properties of materials. One advantage of this approach is reversibility, allowing for recovery of soluble species. There are a variety of species including anthracenes, uracils, and cinnamic acids [107–109]. Notably, Yu et al. used a coumarin motif to cross-link a DPP-based conjugated polymer, as shown in Figure 12A [110].

Cross-linking was confirmed through chemical etching in chloroform, as well as changes in the UV-vis absorption region at 264–366 nm after exposure to 365 nm light, which corresponds to the optical spectral signature of coumarin. Furthermore, through UV-vis spectroscopy, it was shown that the photodimerization was reversible through exposure to 254-nm light. Additionally, GIWAXS studies before and after cross-linking revealed identical morphologies, demonstrating that the coumarin-based crosslinking did not significantly alter the solid-state structure. When integrated into OFETs, the

A)

● **Good Healing Ability**
● **Improvement of Charge Transport Thermal Stability**

B)

C)

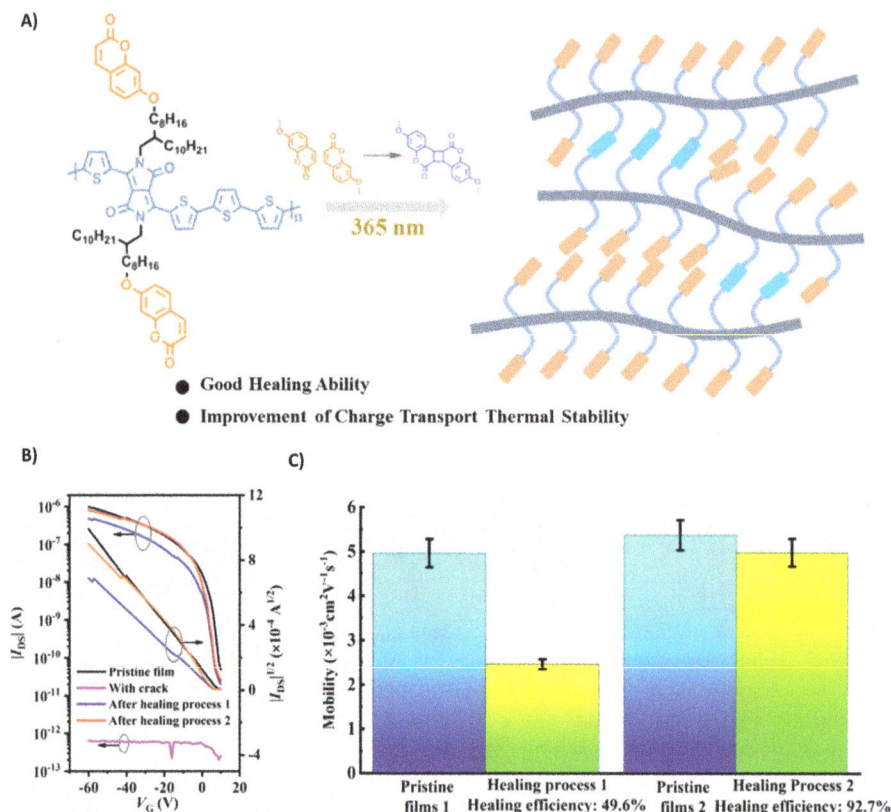

Figure 12: A) The chemical structure of **PDPP4T-DCM** with photo-cross-linkable coumarins in the side chains and illustration of the cross-linking of polymer chains through the photo-induced dimerization of coumarins. B) The representative transfer curve during the healing processes; C) Comparison of hole mobilities of the pristine thin films and the healed thin films. Adapted with permission from reference [110]. Copyright Wiley, 2022.

cross-linking was shown to enhance the hole mobility from 2.86×10^{-3} cm^2 V^{-1} s^{-1} to 2.86×10^{-2} cm^2 V^{-1} s^{-1}. Finally, upon introducing cracks to the thin films and healing using 365-nm light and solvent vapor, 92.7% recovery of charge mobility was observed, indicating a highly efficient cross-linking for self-healing electronics.

Recently, Nyayachavadi et al. developed a photo-cross-linking strategy through the incorporation of reactive 1,3-butadiyne motifs within the side chains of iI-based donor-acceptor conjugated polymers capable of undergoing a topochemical polymerization to form polydiacetylenes [111]. Unique to this strategy is the additive free nature of the cross-linking, as well as the formation of extended conjugated pathways across the alkyl solubilizing regions of the conjugated polymer, having novel implications in organic electronics.

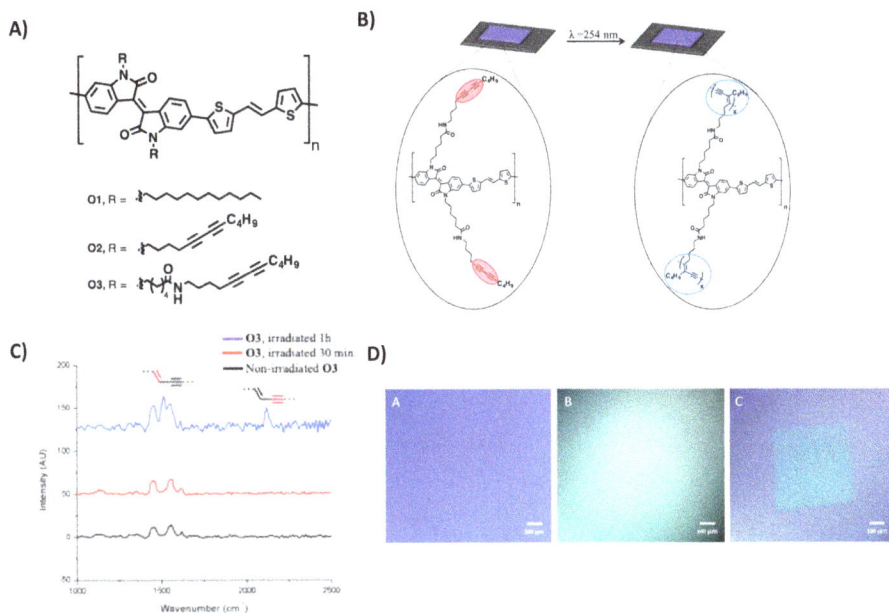

Figure 13: A) Structure of isoindigo-based conjugated polymers containing butadiyne side chains. B) Representative thin-film cross-linking methodology of topochemical polymerization of butadiyne side chains. C) Raman Spectra of **O3** after UV irradiation. D) Thin-film patterning of **O3** via negative photoresist lithography. Adapted with permission from reference [111]. Copyright American Chemical Society, 2019.

Three iI-based polymers were designed, with **O1** acting as a reference material, **O2** containing butadiyne side chains, and **O3** containing butadiyne amide side chains as it has been demonstrated in the literature that guiding noncovalent interactions such as hydrogen bonding are necessary for the topochemical polymerization to occur. Among the three polymers, only **O3** was shown to be capable of photo-cross-linking, as confirmed through Raman spectroscopy as seen in Figure 13C. Additionally, morphological characterizations revealed that the cross-linking did not alter the nanostructure or the crystallinity of the conjugated thin films, indicating good preservation of solid-state structure. Furthermore, transient absorption laser spectroscopy experiments revealed that the photo irradiation cross-linking treatment did not degrade the iI-based polymer backbone and that the formation of PDA within the side chains were independent conjugation pathways. Finally, as shown on Figure 13D, photolithography was shown to be possible directly on the thin films of **O3**, ultimately concluding that this approach was suitable for patternable organic electronics, while maintaining material electronic integrity. This strategy was expanded by Otep et al., who sought to incorporate the PDA within the conjugated backbone of polythiophene and thienothiophene-based conjugated polymer thin films to develop 2D-conjugated polymer thin films as represented in Figure 14A [112].

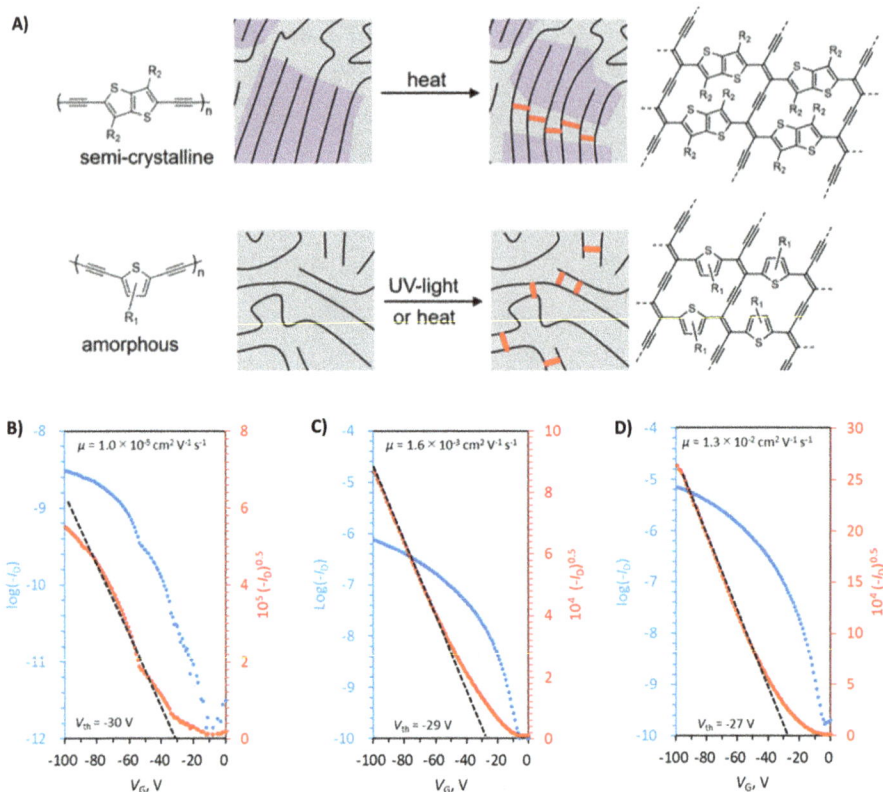

Figure 14: A) Representative crosslinking of **PDETT** (top) and **PDET** (bottom). B) Transfer curve of **PDET**. C) Transfer curve of **PDETT** before thermal annealing. D) Transfer curve of **PDETT** after thermal annealing. Reproduced with permission from reference [112]. Copyright American Chemical Society, 2021.

Upon confirmation of cross-linking through Raman spectroscopy, GIWAXS analysis revealed no significant change in morphology or crystallinity upon formation of PDA cross-links. Notably, it was found that **PDETT** could only be cross-linked thermally as compared to **PDET**, which was attributed to its more crystalline nature and requiring thermal input for chain reorganization and topochemical facilitation. Upon integration into OFETs, it was found that cross-linking of **PDET** was detrimental to the charge transport and yielded no working devices, while cross-linking of **PDETT** showed an improvement in charge mobility by an order of magnitude, indicating that the formation of PDAs directly within the conjugated backbone of polymers result in different electronic properties depending on chemical structure. Furthermore, these results indicated that further work and systematic studies need to be performed to understand the influence of PDA in organic electronics when combined with other electroactive systems.

4.2 Chemical cross-linking

In comparison to photo-cross-linking approaches, there is a vested interest in developing cross-linking strategies for organic semiconductors without the use of photo-initiators, primarily due to the mitigation of degradation that can occur when exposing conjugated materials to light for extended periods of time [60]. While these strategies are not as numerous as photo-initiated ones, there is still an active desire to expand upon them.

Wang et al. reported a side-chain approach to cross-linking by designing DPP-based conjugated polymers containing varying ratios of terminal vinyl side chains that could be covalently cross-linked through the addition of poly(dimethylsiloxane) (PDMS) oligomers via Karstedt's catalyst [113] (Figure 15). PDMS was selected as a cross-linker because it is a well-known elastomeric material with a Young's moduli range reported to be between 360 and 870 KPa, making it an ideal platform for materials that target stretchable applications; by chemically binding adjacent high electronic performance DPP-based conjugated polymers with PDMS, it would be possible to create stretchable electronics that maintain good charge-carrier mobility.

Upon confirmation that the PDMS oligomers were incorporated within the side chains of the DPP polymers via attenuated total-reflectance IR spectroscopy, 20 % (**20DPPTTECx**) was focused on for further analysis due to their maintenance of good

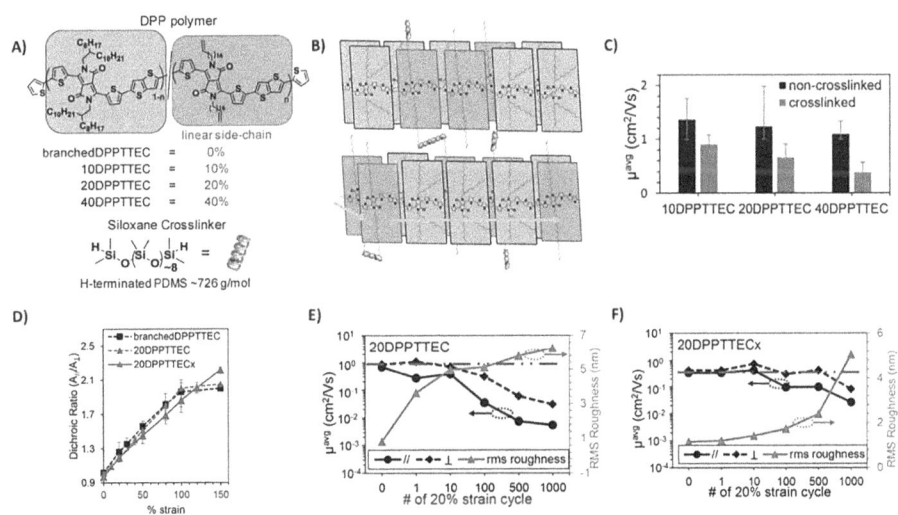

Figure 15: A) Chemical structure of DPP-based random copolymer containing cross-linkable linear side chains and a linear, H-terminated PDMS cross-linker. B) Schematic representation of the covalently cross-linked film. C) Average field-effect mobility of cross-linked polymers compared with uncross-linked polymers. D) Dichroic ratio of **20DPPTTECx** with uncross-linked counterparts and control polymer branched-**DPPTTEC** over a range of strains. E) RMS roughness and average mobility of **20DPPTTEC** and F) **20DPPTTECx** with increasing cycles of 20 % strain. Adapted with permission from reference [113]. Copyright Wiley, 2016.

charge-carrier mobility upon cross-linking in comparison to the 40 % (**40DPPTTECx**), which showed a drastic decrease in electronic performance that was attributed to the plasticizing effect of the PDMS species as confirmed through GIXRD. Conversely, despite the 10 % (**10DPPTTECx**) showing good charge mobility even after cross-linking, the cross-linking density was observed to be too low to have any change in solubility, thus making purification of unreacted materials on the thin-film substrates via chemical washing inaccessible and demonstrating that careful optimization of cross-linking density is required for generating processable, high-performance materials. To test the effects of cross-linking on the mechanical properties, the dichroic ratio of **20DPPTTEC** and **20DPPTTECx** was evaluated in comparison to the reference polymer that has only branched side chains, and it was shown that, even at 150 % strain, the cross-linked polymer showed a linear increase in dichroic ratio in comparison to the uncross-linked and reference species plateaued at 100 % elastic strain, indicative of the formation of cracks confirmed via optical microscopy. Furthermore, the tensile modulus was measured through mechanical buckling, and it was observed that **20DPPTTECx** had a measured Young's modulus $\approx$200 MPa lower than the uncross-linked and reference species, further demonstrating that the cross-linked film had improved ductility and elasticity. Finally, the cross-linked film showed a stable charge mobility of $\approx$0.4 cm^2 V^{-1} s^{-1} up to 500 strain-and-release cycles of 20 % strain perpendicular to charge transport, whereas a sharp decrease in mobility was observed after ten strain-and-release cycles for **20DPPTTEC**, further demonstrating that the PDMS-based cross-linking improves the mechanical properties of organic semiconductors, while helping maintain good electronic performance.

Notably, Hammer et al. devised a reversible covalent cross-linking strategy for conjugated polymers by designing poly(3-hexylthiophene) (P3HT) that possessed terminal thiol side chains capable of spontaneously cross-linking through oxidative formation of disulfide bridges [114].

Upon synthesis, the thioacetate groups were deprotected using an iron(III) chloride oxidation agent, upon which the exposed terminal thiol motifs would form disulfide bridges. Notably, to induce aggregation to form nanowires, CH$_2$Cl$_2$ was added as an antisolvent, which was preliminarily confirmed through UV-Vis spectra by an observable red shift (Figure 16C). Furthermore, this process was shown to be reversible through the addition of dithiothreitol, which would reduce the polymers back to the terminal thiol side chains, thus uncross-linking them and recovering the original optical properties. As a follow up, transmission electron microscopy (TEM) images of the cross-linked nanowires that were drop casted showed bundles with individual structures having widths ~20–25 nm and lengths ~0.2–1 μm (Figure 16D). Furthermore, the cross-linked nanowires showed stability to chemical etching in a variety of tested solvents (CHCl$_3$, chlorobenzene, and toluene) that would readily dissolve uncross-linked nanowires. However, upon integration into field-effect transistors, no notable difference in charge mobility was observed between the cross-linked and uncross-linked polymer species, indicating that this methodology doesn't hinder electronic performance.

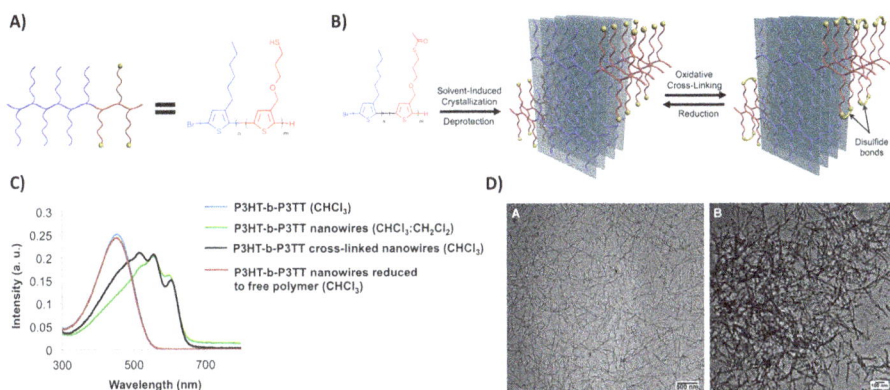

Figure 16: A) Chemical structure of **P3HT-b-P3TT** species. B) Solution-driven assembly represented by stacked P3HT chains and reversible cross-linking of thiol-functionalized blocks. C) UV–vis spectra. **P3HT-b-P3TT** solvated (CHCl₃), nanowires (1:7 CHCl₃:CH2Cl₂), cross-linked nanowires (CHCl₃), reduction to solvated polymer (CHCl₃). D) TEM images of (A) **P3HT-b-P3TT** nanowires (drop cast from 1:7 CHCl₃:CH₂Cl₂); and (B) cross-linked **P3HT-b-P3ST** nanowires (≥6 mM). Adapted with permission from reference [114]. Copyright American Chemical Society, 2014.

Overall, these results demonstrated that this approach was a potential avenue to stabilize desired polymer nanostructures towards specific applications. Lin et al. reported the preparation of an epoxy-functionalized polyfluorene-based conjugated polymer (labeled "PFEX" as shown in Figure 17A) that could be thermally cross-linked upon the addition of a secondary amine additive (TAA) which could cross-link adjacent polymer chains through the highly efficient epoxy-amine reaction as the cathode interlayer for inverted polymer solar cells (IPSC), as per Figure 17B [115].

The cross-linked films were prepared by spin-coating a blend of PFEX and TAA and then annealing at 80 °C for 10 minutes. To control the crosslinking density, varying concentrations of TAA were used, which was qualitatively confirmed through IR and quantitively calculated using XPS. To test the solvent resistant capabilities of the cross-linked films, the UV–vis was investigated as per Figure 17C. The intensities of the uncross-linked film (pure PFEX) washed by chloroform showed very large decline. However, crosslinked films **PFEX-TAA$_{0.022}$**, **PFEX-TAA$_{0.047}$**, **PFEX-TAA$_{0.058}$** with TAA weight ratios of 0.022, 0.047, and 0.058 can maintain almost 95 % absorption intensities, indicating that 95 % of the cross-linked films are stable and unwashed. A further increase of TAA weight ratio to 0.076, 0.12, and 0.19 in the cross-linked thin film (**PFEX-TAA$_{0.076}$**, **PFEX-TAA$_{0.12}$**, **PFEX-TAA$_{0.19}$**) maintained nearly 100 % of the thin film, indicating complete cross-linking. When integrated into inverted polymer solar cells (IPSC), a positive trend was observed with increasing cross-linking density and power conversion efficiency, notably, with **PFEX-TAA$_{0.19}$** reported at 9.4 % in comparison to the uncross-linked PFEX at 6.8 %, indicating that cross-linking helped enhanced the electronic properties of the devices.

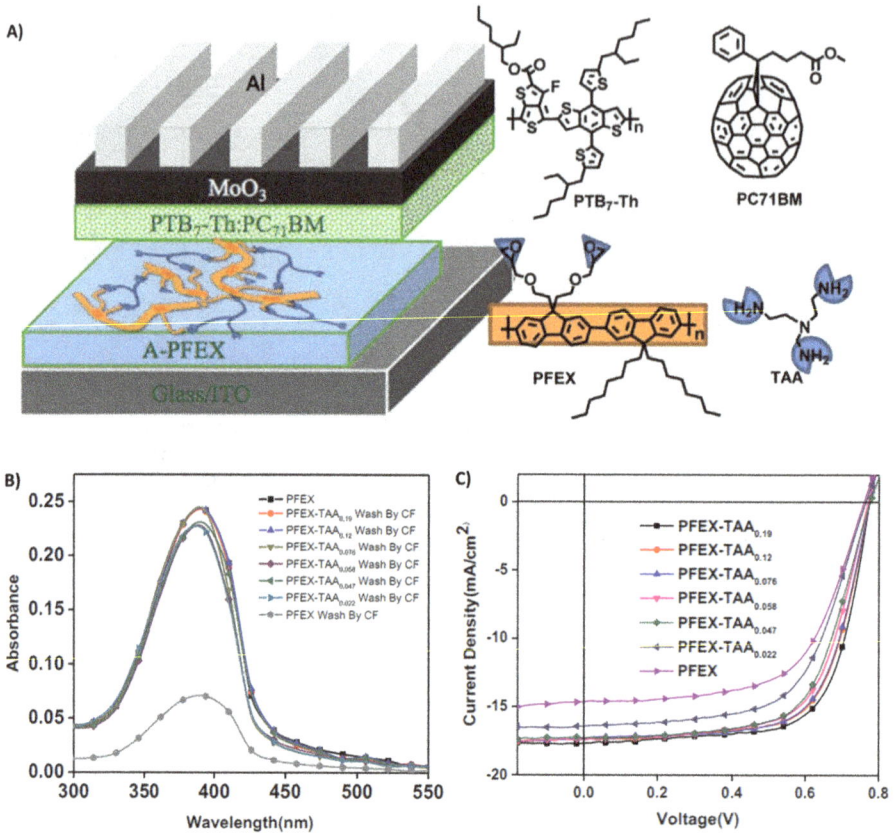

Figure 17: A) Chemical structure of PFEX. B) The device architecture of the IPSC and the molecular structures of PTB7-Th, $PC_{71}BM$, PFEX, and TAA. C) The UV–vis absorption spectra of cross-linked and uncrosslinked PFEX film before and after being washed by chloroform. D) Current density–voltage (J-V) characteristics of PTB7-Th: $PC_{71}BM$-based inverted polymer solar cell. Adapted with permission from reference [115]. Copyright Elsevier, 2017.

5 Conclusion

Organic conjugated polymers have demonstrated themselves to be a powerful class of candidate materials for developing next-generation technologies since they can address many key concerns and desires with regards to the current technological status quo, namely, sustainability through clean manufacturing processes and greener starting materials, targeted thermomechanical and biocompatible application through rational chemical design, and lower overall costs through inexpensive source materials and less resource processing. Despite these potential advantages, conjugated polymers still face issues such as long-term electronic and morphological stability, as well as a decrease in desirable thermomechanical characteristics over time. To address these issues, the application and design of cross-linking strategies within conjugated polymeric systems has shown to be a powerful approach for not only stabilizing but also improving the thermomechanical properties of systems, while also being an additional avenue to tune both the solubility and optoelectronic properties of materials. Because cross-linking was able to expand the scope and application of traditional, nonconjugated polymers to include industrial applications, starting with the vulcanization of rubber for car tires to countless materials that make everyday life possible, the realm of electroactive materials consequently is undergoing a similar scientific revolution where the exploration of new design strategies for conjugated polymers through cross-linking is showing great promise for designing new electronic technologies. It is without a doubt that, as research continues to discover, refine, and optimize new strategies for improving the properties of conjugated polymers, the application of cross-linking and its crossover from research labs to industrial and biomedical sectors is becoming inevitable.

Bibliography

[1]   Namazi, H. Polymers in our daily life. *BioImpacts* **2017**, *7*, 73–74.
[2]   White, J. L.; Spruiell, J. E. The specification of orientation and its development in polymer processing. *Polym. Eng. Sci.* **1983**, *23*, 247–256.
[3]   Vroman, I.; Tighzert, L. Biodegradable polymers. *Materials (Basel)* **2009**, *2*, 307–344.
[4]   Kannurpatti, A. R.; Anderson, K. J.; Anseth, J. W.; Bowman, C. N. Use of "living" radical polymerizations to study the structural evolution and properties of highly crosslinked polymer networks. *J. Polym. Sci., Part B, Polym. Phys.* **1997**, *35*, 2297–2307.
[5]   Escobedo, F. A.; De Pablo, J. J. Monte Carlo simulation of branched and crosslinked polymers. *J. Chem. Phys.* **1996**, *104*, 4788–4801.
[6]   Kaiser, T. Highly crosslinked polymers. *Prog. Polym. Sci.* **1989**, *14*, 373–450.
[7]   Stutz, H.; Illers, K.-H.; Mertes, J. A generalized theory for the glass transition temperature of crosslinked and uncrosslinked polymers. *J. Polym. Sci., Part B, Polym. Phys.* **1990**, *28*, 1483–1498.
[8]   Zhang, S.; Alesadi, A.; Selivanova, M.; Cao, Z.; Qian, Z.; Luo, S.; Galuska, L.; Teh, C.; Ocheje, M. U.; Mason, G. T.; St. Onge, P. B. J.; Zhou, D.; Rondeau-Gagné, S.; Xia, W.; Gu, X. Toward the Prediction and control of glass transition temperature for donor–acceptor polymers. *Adv. Funct. Mater.* **2020**, *30*, 2002221.
[9]   Bandyopadhyay, A.; Valavala, P. K.; Clancy, T. C.; Wise, K. E.; Odegard, G. M. Molecular modeling of crosslinked epoxy polymers: the effect of crosslink density on thermomechanical properties. *Polymer* **2011**, *52*, 2445–2452.
[10] Shen, J.; Lin, X.; Liu, J.; Li, X. Effects of cross-link density and distribution on static and dynamic properties of chemically cross-linked polymers. *Macromolecules* **2019**, *52*, 121–134.
[11] Phadke, A. A.; Bhattacharya, A. K.; Chakraborty, S. K.; De, S. K. Studies of vulcanization of reclaimed rubber. *Rubber Chem. Technol.* **1983**, *56*, 726–736.
[12] Osada, Y.; Gong, J.-P. Soft and wet materials: polymer gels. *Adv. Mater.* **1998**, *10*, 827–837.
[13] Pathania, A.; Arya, R. K.; Ahuja, S. Crosslinked polymeric coatings: preparation, characterization, and diffusion studies. *Prog. Org. Coat.* textbf2017, *105*, 149–162.
[14] Matos-Pérez, C. R.; White, J. D.; Wilker, J. J. Polymer composition and substrate influences on the adhesive bonding of a biomimetic, cross-linking polymer. *J. Am. Chem. Soc.* **2012**, *134*, 9498–9505.

[15] Mavila, S.; Eivgi, O.; Berkovich, I.; Lemcoff, N. G. Intramolecular cross-linking methodologies for the synthesis of polymer nanoparticles. *Chem. Rev.* **2016**, *116*, 878–961.

[16] Bicerano, J.; Sammler, R. L.; Carriere, C. J.; Seitz, J. T. Correlation between glass transition temperature and chain structure for randomly crosslinked high polymers. *J. Polym. Sci., Part B, Polym. Phys.* **1996**, *34*, 2247–2259.

[17] Yamamoto, O.; Kambe, H. Thermal conductivity of cross-linked polymers. A comparison between measured and calculated thermal conductivities. *Polym. J.* **1971**, *2*, 623–628.

[18] Masoumi, H.; Ghaemi, A.; Gilani, H. G. Evaluation of hyper-cross-linked polymers performances in the removal of hazardous heavy metal ions: a review. *Sep. Purif. Technol.* **2021**, *260*, 118221.

[19] Tsyurupa, M. P.; Andreeva, A. I.; Davankov, V. A. On factors determining the swelling ability of crosslinked polymers. *Angew. Makromol. Chem.* **1978**, *70*, 179–187.

[20] Asmussen, E.; Peutzfeldt, A. Influence of selected components on crosslink density in polymer structures. *Eur. J. Oral Sci.* **2001**, *109*, 282–285.

[21] Tung, C. M.; Dynes, P. J. Relationship between viscoelastic properties and gelation in thermosetting systems. *J. Appl. Polym. Sci.* **1982**, *27*, 569–574.

[22] Hajikhani, A.; Wriggers, P.; Marino, M. Chemo-mechanical modelling of swelling and crosslinking reaction kinetics in alginate hydrogels: a novel theory and its numerical implementation. *J. Mech. Phys. Solids* **2021**, *153*, 104476.

[23] Sienkiewicz, A.; Krasucka, P.; Charmas, B.; Stefaniak, W.; Goworek, J. Swelling effects in cross-linked polymers by thermogravimetry. *J. Therm. Anal. Calorim.* **2017**, *130*, 85–93.

[24] Anseth, K. S.; Bowman, C. N. Kinetic gelation model predictions of crosslinked polymer network microstructure. *Chem. Eng. Sci.* **1994**, *49*, 2207–2217.

[25] An, S. Y.; Arunbabu, D.; Noh, S. M.; Song, Y. K.; Oh, J. K. Recent strategies to develop self-healable crosslinked polymeric networks. *Chem. Commun.* **2015**, *51*, 13058–13070.

[26] Hunger, K.; Schmeling, N.; Jeazet, H. B. T.; Janiak, C.; Staudt, C.; Kleinermanns, K. Investigation of cross-linked and additive containing polymer materials for membranes with improved performance in pervaporation and gas separation. *Membranes (Basel)* **2012**, *2*, 727–763.

[27] Fox, T. G.; Loshaek, S. Influence of molecular weight and degree of crosslinking on the specific volume and glass temperature of polymers. *J. Polym. Sci.* **1955**, *15*, 371–390.

[28] Hesekamp, D.; Broecker, H. C.; Pahl, M. H. Chemo-rheology of cross-linking polymers. *Chem. Eng. Technol.* **1998**, *21*, 149–153.

[29] Wang, Z.; Volinsky, A. A.; Gallant, N. D. Crosslinking effect on polydimethylsiloxane elastic modulus measured by custom-built compression instrument. *J. Appl. Polym. Sci.* **2014**, *131*, 1–4.

[30] Xie, T.; Rousseau, I. A. Facile tailoring of thermal transition temperatures of epoxy shape memory polymers. *Polymer* **2009**, *50*, 1852–1856.

[31] Denizli, B. K.; Can, H. K.; Rzaev, Z. M. O.; Guner, A. Preparation conditions and swelling equilibria of Dextran hydrogels prepared by some crosslinking agents. *Polymer* **2004**, *45*, 6431–6435.

[32] Ahn, J. H.; Jang, J. E.; Oh, C. G.; Ihm, S. K.; Cortez, J.; Sherrington, D. C. Rapid generation and control of microporosity, bimodal pore size distribution, and surface area in Davankov-type hyper-cross-linked resins. *Macromolecules* **2006**, *39*, 627–632.

[33] An, M.; Demir, B.; Wan, X.; Meng, H.; Yang, N.; Walsh, T. R. Predictions of thermo-mechanical properties of cross-linked polyacrylamide hydrogels using molecular simulations. *Adv. Theory Simul.* **2019**, *2*, 1–13.

[34] Besghini, D.; Mauri, M.; Simonutti, R. Time domain NMR in polymer science: from the laboratory to the industry. *Appl. Sci.* **2019**, *9*.

[35] Perruchot, C.; Abel, M. L.; Watts, J. F.; Lowe, C.; Maxted, J. T.; White, R. G. High-resolution XPS study of crosslinking and segregation phenomena in hexamethoxymethyl melamine-polyester resins. *Surf. Interface Anal.* **2002**, *34*, 570–574.

[36] Nielsen, L. E. Polymer reviews cross-linking – Effect on physical properties of polymers. *J. Macromol. Sci., Part C, Polym. Rev.* **2008**, *3*, 69–103.

[37] Tillet, G.; Boutevin, B.; Ameduri, B. Chemical reactions of polymer crosslinking and post-crosslinking at room and medium temperature. *Prog. Polym. Sci.* **2011**, *36*, 191–217.

[38] Gubbi, J.; Buyya, R.; Marusic, S.; Palaniswami, M. Internet of Things (IoT): a vision, architectural elements, and future directions. *Future Gener. Comput. Syst.* **2013**, *29*, 1645–1660.

[39] Venema, L. Silicon electronics and beyond. *Nature* **2011**, *479*, 309.

[40] Forrest, S. R. The path to ubiquitous and low-cost organic electronic appliances on plastic. *Nature* **2004**, *428*, 911–918.

[41] Keyes, R. W. Fundamental Limits of silicon. *Proc. IEEE* **2001**, *89*, 227–239.

[42] Calderón Gómez, D. Technological capital and digital divide among young people: an intersectional approach. *J. Youth Stud.* **2019**, *22*, 941–958.

[43] Margalit, N.; Xiang, C.; Bowers, S. M.; Bjorlin, A.; Blum, R.; Bowers, J. E. Perspective on the future of silicon photonics and electronics. *Appl. Phys. Lett.* **2021**, *118*, 220501.

[44] O'connor, T. F.; Rajan, K. M.; Printz, A. D.; Lipomi, D. J. Toward organic electronics with properties inspired by biological tissue. *J. Mater. Chem. B* **2015**, *3*, 4947–4952.

[45] Heremans, P.; Tripathi, A. K.; de Jamblinne de Meux, A.; Smits, E. C. P.; Hou, B.; Pourtois, G.; Gelinck, G. H. Mechanical and electronic properties of thin-film transistors on plastic, and their integration in flexible electronic applications. *Adv. Mater.* **2016**, *28*, 4266–4282.

[46] Marrocchi, A.; Trombettoni, V.; Sciosci, D.; Campana, F.; Vaccaro, L. *Key trends in sustainable approaches to the synthesis of semiconducting polymers*, 2nd ed.; Elsevier Ltd., 2019.

[47] Shirakawa, H. The discovery of polyacetylene film: the dawning of an era of conducting polymers (Nobel lecture). *Angew. Chem., Int. Ed.* **2001**, *40*, 2574–2580.

[48] Wang, M.; Baek, P.; Akbarinejad, A.; Barker, D.; Travas-Sejdic, J. Conjugated polymers and composites for stretchable organic electronics. *J. Mater. Chem. C* **2019**, *7*, 5534–5552.

[49] Nayana, V.; Kandasubramanian, B. Polycarbazole and its derivatives: progress, synthesis, and applications. *J. Polym. Res.* **2020**, *27*, 29–32.

[50] Mike, J. F.; Lutkenhaus, J. L. Recent advances in conjugated polymer energy storage. *J. Polym. Sci., Part B, Polym. Phys.* **2013**, *51*, 468–480.

[51] Clingerman, M. L.; King, J. A.; Schulz, K. H.; Meyers, J. D. Evaluation of electrical conductivity models for conductive polymer composites. *J. Appl. Polym. Sci.* **2002**, *83*, 1341–1356.

[52] Du, W.; Ohayon, D.; Combe, C.; Mottier, L.; Maria, I. P.; Ashraf, R. S.; Fiumelli, H.; Inal, S.; McCulloch, I. Improving the compatibility of diketopyrrolopyrrole semiconducting polymers for biological interfacing by lysine attachment. *Chem. Mater.* **2018**, *30*, 6164–6172.

[53] Reid, D. R.; Jackson, N. E.; Bourque, A. J.; Snyder, C. R.; Jones, R. L.; De Pablo, J. J. Aggregation and solubility of a model conjugated donor-acceptor polymer. *J. Phys. Chem. Lett.* **2018**, *9*, 4802–4807.

[54] Kim, M.; Ryu, S. U.; Park, S. A.; Choi, K.; Kim, T.; Chung, D.; Park, T. Donor–acceptor-conjugated polymer for high-performance organic field-effect transistors: a progress report. *Adv. Funct. Mater.* **2020**, *30*, 1–25.

[55] Ocheje, M. U.; Charron, B. P.; Nyayachavadi, A.; Rondeau-Gagné, S. Stretchable electronics: Recent progress in the preparation of stretchable and self-healing semiconducting conjugated polymers. *Flex. Print. Electron.* **2017**, *2*, 043002.

[56] St. Onge, P. B. J.; Ocheje, M. U.; Selivanova, M.; Rondeau-Gagné, S. Recent advances in mechanically robust and stretchable bulk heterojunction polymer solar cells. *Chem. Rec.* **2019**, *19*, 1008.

[57] Chen, A. X.; Hilgar, J. D.; Samoylov, A. A.; Pazhankave, S. S.; Bunch, J. A.; Choudhary, K.; Esparza, G. L.; Lim, A.; Luo, X.; Chen, H.; Runser, R.; McCulloch, I.; Mei, J.; Hoover, C.; Printz, A. D.; Romero, N. A.; Lipomi, D. J. Increasing the strength, hardness, and survivability of semiconducting polymers by crosslinking. *Adv. Mater. Interfaces* **2023**, *10*.

[58] Kahle, F. J.; Saller, C.; Köhler, A.; Strohriegl, P. Crosslinked semiconductor polymers for photovoltaic applications. *Adv. Energy Mater.* **2017**, *7*, 1–10.

[59] Wang, X.; Deng, J.; Duan, X.; Liu, D.; Guo, J.; Liu, P. Crosslinked polyaniline nanorods with improved electrochemical performance as electrode material for supercapacitors. *J. Mater. Chem. A* **2014**, *2*, 12323–12329.

[60] Freudenberg, J.; Jänsch, D.; Hinkel, F.; Bunz, U. H. F. Immobilization strategies for organic semiconducting conjugated polymers. *Chem. Rev.* **2018**, *118*, 5598–5689.

[61] Zeglio, E.; Rutz, A. L.; Winkler, T. E.; Malliaras, G. G.; Herland, A. Conjugated polymers for assessing and controlling biological functions. *Adv. Mater.* **2019**, *31*.

[62] Zhao, C.; Chen, Z.; Shi, R.; Yang, X.; Zhang, T. Recent advances in conjugated polymers for visible-light-driven water splitting. *Adv. Mater.* **2020**, *32*, 1–52.

[63] Leclère, P.; Surin, M.; Brocorens, P.; Cavallini, M.; Biscarini, F.; Lazzaroni, R. Supramolecular assembly of conjugated polymers: from molecular engineering to solid-state properties. *Mater. Sci. Eng., R Rep.* **2006**, *55*, 1–56.

[64] Zhu, H.; Luo, W.; Ciesielski, P. N.; Fang, Z.; Zhu, J. Y.; Henriksson, G.; Himmel, M. E.; Hu, L. Wood-derived materials for green electronics, biological devices, and energy applications. *Chem. Rev.* **2016**, *116*, 9305–9374.

[65] Jeffrey, G. A. Hydrogen-bonding: an update. *Crystallogr. Rev.* **2003**, *9*, 135–176.

[66] Stupp, S. I.; Palmer, L. C. Supramolecular chemistry and self-assembly in organic materials design. *Chem. Mater.* **2014**, *26*, 507–518.

[67] Mullin, W. J.; Sharber, S. A.; Thomas, S. W. Optimizing the self-assembly of conjugated polymers and small molecules through structurally programmed non-covalent control. *J. Polym. Sci.* **2021**, *59*, 1643–1663.

[68] Shi, X.; Bao, W. Hydrogen-bonded conjugated materials and their application in organic field-effect transistors. *Front. Chem.* **2021**, *9*, 1–6.

[69] Głowacki, E. D.; Irimia-Vladu, M.; Bauer, S.; Sariciftci, N. S. Hydrogen-bonds in molecular solids – From biological systems to organic electronics. *J. Mater. Chem. B* **2013**, *1*, 3742–3753.

[70] González-Rodríguez, D.; Schenning, A. P. H. J. Hydrogen-bonded supramolecular π-functional materials. *Chem. Mater.* **2011**, *23*, 310–325.

[71] Mu, Y.; Sun, Q.; Wan, X. Impact of polymer chemistry on the application of polyurethane / ureas in organic thin film transistors. *RSC Appl. Polym.* **2023**, *1*, 190–203.

[72] Oh, J. Y.; Rondeau-Gagné, S.; Chiu, Y. C.; Chortos, A.; Lissel, F.; Wang, G. J. N.; Schroeder, B. C.; Kurosawa, T.; Lopez, J.; Katsumata, T.; et al. Intrinsically stretchable and healable semiconducting polymer for organic transistors. *Nature* **2016**, *539*, 411–415.

[73] Ocheje, M. U.; Charron, B. P.; Cheng, Y. H.; Chuang, C. H.; Soldera, A.; Chiu, Y. C.; Rondeau-Gagné, S. Amide-containing alkyl chains in conjugated polymers: effect on self-assembly and electronic properties. *Macromolecules* **2018**, *51*, 1336–1344.

[74] Ye, L.; Pankow, R. M.; Horikawa, M.; Melenbrink, E. L.; Liu, K.; Thompson, B. C. Green-solvent-processed amide-functionalized conjugated polymers prepared via direct arylation polymerization (DArP). *Macromolecules* **2019**, *52*, 9383–9388.

[75] Mooney, M.; Wang, Y.; Nyayachavadi, A.; Zhang, S.; Gu, X.; Rondeau-Gagné, S. Enhancing the solubility of semiconducting polymers in eco-friendly solvents with carbohydrate-containing side chains. *ACS Appl. Mater. Interfaces* **2021**, *13*, 25175–25185.

[76] Sabury, S.; Adams, T. J.; Kocherga, M.; Kilbey, S. M.; Walter, M. G. Synthesis and optoelectronic properties of benzodithiophene-based conjugated polymers with hydrogen bonding nucleobase side chain functionality. *Polym. Chem.* **2020**, *11*, 5735–5749.

[77] Xu, Y.; Zhang, F.; Feng, X. Patterning of conjugated polymers for organic optoelectronic devices. *Small* **2011**, *7*, 1338–1360.

[78] Yang, K.; He, T.; Chen, X.; Cheng, S. Z. D.; Zhu, Y. Patternable conjugated polymers with latent hydrogen-bonding on the main chain. *Macromolecules* **2014**, *47*, 8479–8486.

[79] Guo, Z. H.; Ai, N.; McBroom, C. R.; Yuan, T.; Lin, Y. H.; Roders, M.; Zhu, C.; Ayzner, A. L.; Pei, J.; Fang, L. A side-chain engineering approach to solvent-resistant semiconducting polymer thin films. *Polym. Chem.* **2016**, *7*, 648–655.

[80] Ji, J.; Zhou, D.; Tang, Y.; Deng, P.; Guo, Z.; Zhan, H.; Yu, Y.; Lei, Y. Partially removing long branched alkyl side chains of regioregular conjugated backbone based diketopyrrolopyrrole polymer for improving field-effect mobility. *J. Mater. Chem. C* **2018**, *6*, 13325–13330.

[81]   Smith, Z. C.; Meyer, D. M.; Simon, M. G.; Staii, C.; Shukla, D.; Thomas, S. W. Thiophene-based conjugated polymers with photolabile solubilizing side chains. *Macromolecules* **2015**, *48*, 959–966.

[82]   Yao, J.; Yu, C.; Liu, Z.; Luo, H.; Yang, Y.; Zhang, G.; Zhang, D. Significant improvement of semiconducting performance of the diketopyrrolopyrrole-quaterthiophene conjugated polymer through side-chain engineering via hydrogen-bonding. *J. Am. Chem. Soc.* **2016**, *138*, 173–185.

[83]   Yang, Y.; Liu, Z.; Chen, L.; Yao, J.; Lin, G.; Zhang, X.; Zhang, G.; Zhang, D. Conjugated semiconducting polymer with thymine groups in the side chains: charge mobility enhancement and application for selective field-effect transistor sensors toward CO and H 2 S. *Chem. Mater.* **2019**, *31*, 1800–1807.

[84]   Mooney, M.; Wang, Y.; Iakovidis, E.; Gu, X.; Rondeau-Gagné, S. Carbohydrate-containing conjugated polymers: Solvent-resistant materials for greener organic electronics. *ACS Appl. Electron. Mater.* **2022**, *4*, 1381–1390.

[85]   Yang, H.; Yuan, B.; Zhang, X.; Scherman, O. A. Supramolecular chemistry at interfaces: host–guest interactions for fabricating multifunctional biointerfaces. *Acc. Chem. Res.* **2014**, *47*, 2106–2115.

[86]   Pappalardo, A.; Ballistreri, F. P.; Destri, G. L.; Mineo, P. G.; Tomaselli, G. A.; Toscano, R. M.; Trusso Sfrazzetto, G. Supramolecular polymer networks based on calix[5]arene tethered poly(p-phenyleneethynylene). *Macromolecules* **2012**, *45*, 7549–7556.

[87]   Ji, X.; Yao, Y.; Li, J.; Yan, X.; Huang, F. A supramolecular cross-linked conjugated polymer network for multiple fluorescent sensing. *J. Am. Chem. Soc.* **2013**, *135*, 74–77.

[88]   Wu, Y.; Li, H.; Yan, Y.; Shan, X.; Zhao, M.; Zhao, Q.; Liao, X.; Xie, M. Pillararene-containing polymers with tunable conductivity based on host-guest complexations. *ACS Macro Lett.* **2019**, *8*, 1588–1593.

[89]   Fu, T.; Li, Z.; Zhang, Z.; Zhang, X.; Wang, F. Supramolecular cross-linking and gelation of conjugated polycarbazoles via hydrogen bond assisted molecular tweezer/guest complexation. *Macromolecules* **2017**, *50*, 7517–7525.

[90]   Wei, P.; Yan, X.; Huang, F. Supramolecular polymers constructed by orthogonal self-assembly based on host-guest and metal-ligand interactions. *Chem. Soc. Rev.* **2015**, *44*, 815–832.

[91]   Astruc, D. From organotransition-metal chemistry toward molecular electronics: electronic communication between ligand-bridged metals. *Acc. Chem. Res.* **1997**, *30*, 383–391.

[92]   Welterlich, I.; Tieke, B. Conjugated polymer with benzimidazolylpyridine ligands in the side chain: metal ion coordination and coordinative self-assembly into fluorescent ultrathin films. *Macromolecules* **2011**, *44*, 4194–4203.

[93]   Wu, H. C.; Rondeau-Gagne, S. Chiu, Y.-C.; Lissel, F.; To, J. W. F.; Tsao, Y.; Oh, J. Y.; Tang, B.; Chen, W.-C.; Tok, J. B.-H.; Bao, Z. Enhanced charge transport and stability conferred by iron(III)-coordination in a conjugated polymer thin-film transistors. *Adv. Electron. Mater.* **2018**, *4*, 1800239.

[94]   Reddy, N.; Reddy, R.; Jiang, Q. Crosslinking biopolymers for biomedical applications. *Trends Biotechnol.* **2015**, *33*, 362–369.

[95]   Hu, W.; Wang, Z.; Xiao, Y.; Zhang, S.; Wang, J. Advances in crosslinking strategies of biomedical hydrogels. *Biomater. Sci.* **2019**, *7*, 843–855.

[96]   Tian, Y.; Kong, L.; Mao, H.; Shi, J.; Tong, B.; Cai, Z.; Dong, Y. A supramolecular approach for the synthesis of cross-linked ionic polyacetylene network gels. *Mater. Chem. Front.* **2020**, *4*, 645–650.

[97]   Shinde, S.; Gavvalapalli, N. Impact of amine additives on the mechanical properties of hydrogen bonding π-conjugated polymers. *Polymer* **2020**, *204*, 122856.

[98]   Eelkema, R.; Pich, A. Pros and cons: supramolecular or macromolecular: what is best for functional hydrogels with advanced properties? *Adv. Mater.* **2020**, *32*, 1906012.

[99]   Chen, J. T.; Hsu, C. S. Conjugated polymer nanostructures for organic solar cell applications. *Polym. Chem.* **2011**, *2*, 2707–2722.

[100] Wang, J.; Lin, K.; Zhang, K.; Jiang, X. F.; Mahmood, K.; Ying, L.; Huang, F.; Cao, Y. Crosslinkable amino-functionalized conjugated polymer as cathode interlayer for efficient inverted polymer solar cells. *Adv. Energy Mater.* **2016**, *6*, 1–9.

[101] Lin, Y. C.; Chen, C. K.; Chiang, Y. C.; Hung, C. C.; Fu, M. C.; Inagaki, S.; Chueh, C. C.; Higashihara, T.; Chen, W. C. Study on intrinsic stretchability of diketopyrrolopyrrole-based π-conjugated copolymers with poly(acryl amide) side chains for organic field-effect transistors. *ACS Appl. Mater. Interfaces* **2020**, *12*, 33014–33027.

[102] Chen, R.; Wang, X.; Li, X.; Wang, H.; He, M.; Yang, L.; Guo, Q.; Zhang, S.; Zhao, Y.; Li, Y.; Liu, Y.; Wei, D. A comprehensive nano-interpenetrating semiconducting photoresist toward all-photolithography organic electronics. *Sci. Adv.* **2021**, *7*, 1–10.

[103] Kim, M. J.; Lee, M.; Min, H.; Kim, S.; Yang, J.; Kweon, H.; Lee, W.; Kim, D. H.; Choi, J. H.; Ryu, D. Y.; Kang, M. S.; Kim, B. S.; Cho, J. H. Universal three-dimensional crosslinker for all-photopatterned electronics. *Nat. Commun.* **2020**, *11*, 1–11.

[104] Wang, G. J. N.; Zheng, Y.; Zhang, S.; Kang, J.; Wu, H. C.; Gasperini, A.; Zhang, H.; Gu, X.; Bao, Z. Tuning the cross-linker crystallinity of a stretchable polymer semiconductor. *Chem. Mater.* **2019**, *31*, 6465–6475.

[105] Zheng, Y.; Yu, Z.; Zhang, S.; Kong, X.; Michaels, W.; Wang, W.; Chen, G.; Liu, D.; Lai, J. C.; Prine, N.; et al. A molecular design approach towards elastic and multifunctional polymer electronics. *Nat. Commun.* **2021**, *12*, 1–11.

[106] Gao, C.; Shi, D.; Li, C.; Yu, X.; Zhang, X.; Liu, Z.; Zhang, G.; Zhang, D. A dual functional diketopyrrolopyrrole-based conjugated polymer as single component semiconducting photoresist by appending azide groups in the side chains. *Adv. Sci.* **2022**, *9*, 1–8.

[107] Schelkle, K. M.; Bender, M.; Beck, S.; Jeltsch, K. F.; Stolz, S.; Zimmermann, J.; Weitz, R. T.; Pucci, A.; Müllen, K.; Hamburger, M.; Bunz, U. H. F. Photo-cross-linkable polymeric optoelectronics based on the [2 + 2] cycloaddition reaction of cinnamic acid. *Macromolecules* **2016**, *49*, 1518–1522.

[108] Shih, H. K.; Chen, Y. H.; Chu, Y. L.; Cheng, C. C.; Chang, F. C.; Zhu, C. Y.; Kuo, S. W. Photo-crosslinking of pendent uracil units provides supramolecular hole injection/transport conducting polymers for highly efficient light-emitting diodes. *Polymers (Basel)* **2015**, *7*, 804–818.

[109] Ouyang, T.; Guo, X.; Cui, Q.; Zhang, W.; Dong, W.; Fei, T. Conjugated polymer nanoparticles based on anthracene and tetraphenylethene for nitroaromatics detection in aqueous phase. *Chemosensors* **2022**, *10*, 366.

[110] Yu, X.; Li, C.; Gao, C.; Chen, L.; Zhang, X.; Zhang, G.; Zhang, D. Enhancing the healing ability and charge transport thermal stability of a diketopyrrolopyrrole based conjugated polymer by incorporating coumarin groups in the side chains. *J. Polym. Sci.* **2022**, *60*, 517–524.

[111] Nyayachavadi, A.; Langlois, A.; Tahir, M. N.; Billet, B.; Rondeau-Gagné, S. Conjugated polymer with polydiacetylene cross-links through topochemical polymerization of 1,3-butadiyne moieties toward photopatternable thin films. *ACS Appl. Polym. Mater.* **2019**, *1*, 1918–1924.

[112] Otep, S.; Ogita, K.; Yomogita, N.; Motai, K.; Wang, Y.; Tseng, Y. C.; Chueh, C. C.; Hayamizu, Y.; Matsumoto, H.; Ishikawa, K.; Mori, T.; Michinobu, T. Cross-linking of poly(arylenebutadiynylene)s and its effect on charge carrier mobilities in thin-film transistors. *Macromolecules* **2021**, *54*, 4351–4362.

[113] Wang, G. J. N.; Shaw, L.; Xu, J.; Kurosawa, T.; Schroeder, B. C.; Oh, J. Y.; Benight, S. J.; Bao, Z. Inducing elasticity through oligo-siloxane crosslinks for intrinsically stretchable semiconducting polymers. *Adv. Funct. Mater.* **2016**, *26*, 7254–7262.

[114] Hammer, B. A. G.; Reyes-Martinez, M. A.; Bokel, F. A.; Liu, F.; Russell, T. P.; Hayward, R. C.; Briseno, A. L.; Emrick, T. Reversible self cross-linking nanowires from thiol-functionalized polythiophene diblock copolymers. *ACS Appl. Mater. Interfaces* **2014**, *6*, 7705–7711.

[115] Lin, K.; Wang, J.; Hu, Z.; Xu, R.; Liu, J.; Liu, X.; Xu, B.; Huang, F.; Cao, Y. Novel cross-linked films from epoxy-functionalized conjugated polymer and amine based small molecule for the interface engineering of high-efficiency inverted polymer solar cells. *Sol. Energy Mater. Sol. Cells* **2017**, *168*, 22–29.

Edward J. Barron III and Michael D. Bartlett

Soft robotics: a futuristic extension of stretchable electronics technology

Abstract: Multifunctional materials can enable new robotic and electronic technologies through unique combinations of mechanical and functional properties. Soft multifunctional composites, which incorporate functional inclusions such as metals, ceramics, or fluids into soft polymers, have shown exciting properties such as high thermal and electrical conductivity, magnetic response, or stimuli-responsive shape and rigidity tuning. These composites have enabled new forms of electronics and robots that are soft and deformable with capabilities that enable adaptive, responsive, and multifunctional capabilities, reminiscent of capabilities found in biological organisms. In this chapter, we discuss fabrication methods, resulting properties, and applications of several classes of multifunctional composite architectures that integrate functional inclusions with soft polymers. There will be an emphasis on room temperature liquid metals (LM), low melting point alloys (LMPA), and magnetic powders and fluids as inclusions. We will then discuss how integrating synergistic forms of these multifunctional materials can create intriguing possibilities for advanced soft robotic systems.

1 Introduction

Stretchable electronics has been an area of intense research with an emphasis on creating devices that can be mechanically deformed while accomplishing tasks performed by modern devices such as sensors, computers, and smart phones. In contrast to traditional rigid electronics, stretchable electronics allow for greater freedom of application in human–machine interfacing, such as the ability to place sensors on high-deformation regions of the human body, or on delicate tissues for medical applications [1, 2]. Stretchable electronic devices require multifunctional material systems which are capable of achieving functional properties such as electrical and thermal conductivity while maintaining a low mechanical stiffness and high strain limit. Research into these materials has been successful in recent years, which has not only advanced understanding of stretchable electronics technology, but has also enabled emerging technologies such as soft robots. Recent advances in soft robotics have proceeded largely in tandem with soft

Acknowledgement: We acknowledge support from the Office of Naval Research Young Investigator Program (YIP) (N000142112699) and NSF (No. CMMI-2054409).

Edward J. Barron III, Michael D. Bartlett, Mechanical Engineering, Soft Materials and Structures Lab, Virginia Tech, Blacksburg, VA 24061, USA; and Macromolecules Innovation Institute, Virginia Tech, Blacksburg, VA 24061, USA, e-mails: ebarron@vt.edu, mbartlett@vt.edu

https://doi.org/10.1515/9783110757286-003

electronics, and these soft machines have pushed towards bioinspired functions, safe human–machine interaction, and the ability to adapt their properties and function to accomplish diverse tasks [3–6]. As soft robotic research has advanced, the need for new materials has increased [7]. With the evolution of stretchable electronics came the rise of soft multifunctional materials that have enabled significant advances in robots such as self-healing [8–10], high deformability [11], and passive adaptivity [12], which have given rise to a surge of research interest into soft machines [13].

To understand the evolution of soft robots, it is useful to understand the research behind stretchable and flexible electronics and the innovations in soft functional materials. Traditional electronics require functional components such as wiring to transfer power to electronic devices, heat transfer components such as heat sinks that prevent overheating, and often times magnetic materials that improve the efficiency of electromagnetic components. However, the materials that possess the most outstanding electrical, thermal, and magnetic properties are rigid metals and ceramics that are useful in traditional electronics but are a limiting factor in the development of stretchable electronics and soft robots. The ability to produce deformable electronics has largely occurred through the creation of material systems through three techniques: 1) deterministic geometries, such as the patterning of rigid functional materials to globally deform to relatively high strain rates without catastrophic local deformation; 2) intrinsically stretchable materials, such as the chemical synthesis of intrinsically soft and conductive polymers; and 3) soft composites, including the fabrication of composites that utilize functional fillers in highly extensible polymers [14, 15].

1.1 Deterministic geometries

Through deterministic geometries, rigid electronic materials are formulated to take advantage of the comparatively low strains in bending to create stretchable systems [11, 16–18]. Examples of this technique include wrinkled or buckled structures, where thin rigid materials undergo out-of-plane bending to realize large global deformations. These techniques often use thin (<1 μm) functional materials such as single-crystal semiconductors integrated with soft elastomer substrates, allowing the material system to undergo substantial global deformation while maintaining local strains of less than <1 % in the functional material [19, 20]. Rigid materials can be added to a pristine flexible substrate for material systems with high degrees of bending, or added to a prestrained substrate to allow for further axial deformation by leveraging material buckling [21, 22]. In addition to wrinkled geometries, other patterns include the creation of serpentine structures and fractal patterns that allow for tensile stretching and twisting [23–25]. A recent engineering technique used to create deformable electronics has evolved from kirigami, or the art of paper cutting [26, 27]. The technique of applying engineered cuts is scalable, works on a diverse number of materials, and can be applied to rigid functional materials to enable extensibility for flexible electronic components [26–29]. In addition to electronics, kirigami has been utilized to create mutifunctional materials capable of improving soft robots, such as through improving actuation and shape change [30–32].

1.2 Intrinsically stretchable materials

Intrinsically soft and functional polymers have been synthesized for electronics applications. These polymers are typically semiconducting or electrically conductive and are largely synthesized as π-conjugated polymers, often leading to semicrystalline morphologies [33, 34]. These intrinsically stretchable materials are often less deformable than other approaches, but this can be improved by integrating with an elastic substrate or through approaches like confinement that enhance more molecular mobility [34–36]. Another form of conductive polymers are ionic hydrogels and ionogels that utilize electrolytic solutions or ionic liquids [37, 38]. These polymers can be highly elastic, soft, and are capable of the deformation needed for many stretchable electronics applications, however, these intrinsically functional soft polymers often struggle to approach the conductivity of rigid conductors ($\sim 10^{-1}\,\Omega$m), with their strength instead focused on speed of signal transmission ($\sim 10^{-8}$ s) [37, 39].

1.3 Soft composite fabrication

Composites have also been used to create soft functional materials. These composites often rigid functional fillers such as metallic or ceramic nano- and microinclusions including particles [40], flakes [41, 42], wires [43], or nanotubes as inclusions in soft extensible polymers such as gels and elastomers [44]. These materials can be made thermally conductive for heat- transfer applications such as thermal pads and interface materials [45, 46]. Electrically conductive soft composites can be created by utilizing a sufficiently high-volume fraction of conductive filler to reach the percolation threshold, where a continuous electrical pathway is created by the rigid inclusions [47–49]. One downside to this approach is that there is a fundamental trade-off between the mechanical and functional properties of soft composites, where the high volume fractions of functional inclusions often necessary to achieve desirable properties lead to a significant stiffening of the composite and a decrease in the strain limit of the material [50]. Additionally, the functional properties can vary or, in the case of electrical conductivity, disappear eliminated with high deformation due to the strain-dependent microstructure. One recent method utilized to overcome some of these problems has been through the use of functional liquid inclusions. By utilizing fluids instead of rigid particles, soft functional composites can be highly loaded with filler while still maintaining low stiffness and high extensibility [51]. The creation of these solid–fluid composites requires the utilization of fluids with sufficient functionality which has been carried out with room temperature liquid metals and low melting-point alloys that have high thermal

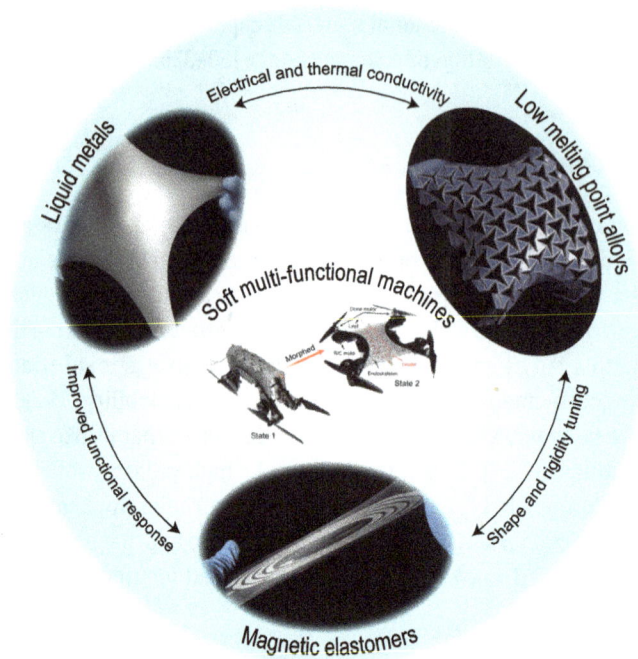

Figure 1: *Soft, functional, and active materials.* Multifunctional composites utilize fillers such as liquid metals, low melting-point alloys, and magnetic particles and fluids to enable deformability, thermal, electrical, and magnetic material properties, and adaptive response for electronics and machines. *Top left image:* A liquid-metal–elastomer composite with high thermal conductivity. Open access article [59]. *Top right image:* A low melting-point alloy composite with shape-morphing capabilities. *Bottom image:* A magnetic elastomer composite with liquid-metal conductive channels. Reprinted (adapted) with permission from [56]. Copyright 2020 American Chemical Society. *Center image:* A multicomponent shape-morphing material integrated with onboard power and control to enable a morphing drone. From [60]. Reprinted with permission from AAAS.

and electrical conductivities [9, 52–55], as well as with magnetic fluids to impart magnetic response [56–58].

The evolution of soft robots and machines has occurred in tandem with advancements in soft functional composites. Recent soft machines have shown incredible properties, such as the ability to self-heal after damage events [9, 60, 61], reconfigure shapes to adapt to different tasks [60, 62], and passively grasp objects of a variety of shapes [63, 64]. Many of these abilities rely on the materials science advances of functional composite materials that make possible functional and adaptive responses on the material level. The precursors to soft robots were actuators that usually used a combination of rigid components and selectively placed compliant regions rather than being developed entirely employing materials [7, 65]. As research into active soft systems evolved, actuators and artificial muscles began to become more commonly researched, with many made out of nearly entirely soft systems such as by those utilizing pneumatic devices to activate rubber-based actuators [66–69]. With the advancement soft composite materials, there is more flexibility in the creation of state-of-the-art soft robots and machines. Pneumatic systems are still heavily utilized in soft robots [70, 71], but in addition to these devices there have been composites that can accomplish material level shape change, stiffness tuning, and actuation via electrical, thermal, and magnetic stimuli [72–74]. The integration of these materials into multicomponent systems significantly increases the potential of machine design and opens pathways for the intelligent design of next-generation robots with unique properties not capable of being achieved by traditional robotic systems.

In this chapter, we will discuss the fabrication, microstructure, and properties of composite materials that enable soft robots and machines (Figure 1). First, there will be a discussion of elastomer composites that utilize room temperature liquid metals for high thermal and electrical conductivities in soft and extensible systems for deformable wiring and efficient heat transfer in soft robots. Then, we will discuss composites with low melting-point alloys (LMPAs), which are defined as solids at room temperature with melting points slightly above room temperature. These materials make possible increases similar to the functional properties seen in liquid metals, while also allow for adaptivity in rigidity and shape. We will then discuss magnetic elastomer composites that utilize magnetic fillers to control magnetic circuits and enable adaptive properties such as rapid shape change and stiffness tuning. Finally, multicomponent composites that utilize a combination of several functional materials in a single system, that can equip soft machines with diverse capabilities will be addressed.

2 Liquid-metal–elastomer composites

2.1 Liquid metal

Metals have long been used in engineering because of their mechanical strength which well support structural applications, plus their functional properties in electronics, such as copper in traditional wiring and aluminum in heat sinks for electronics. However, not all metals exist solely as solids at room temperature. Elements such as mercury ($T_m = -38.8\,°C$), francium ($T_m = 27\,°C$), cesium ($T_m = 28.5\,°C$), gallium ($T_m = 29.8\,°C$), and rubidium ($T_m = 39\,°C$) are metals that exist as liquids at or near room temperature [75–77]. While liquid metals offer intriguing possibilities for soft functional composites, using many of them involve inherent problems. Francium is radioactive and chemically unstable, with a half-life of ~20 min, while cesium and rubidium are explosively reactive which limits their use [78]. Mercury is perhaps the most well-known of these materials due to its heavy usage in devices and materials such as thermometers and dental amalgams [79, 80]. However, mercury's toxicity render it unfeasible for many modern and emerging devices such as wearable and bioelectronics.

Gallium is an intriguing alternative material that liquefies near room temperature, has the low toxicity necessary to be used as contrast agents and in pharmaceuticals, and produces desirable functional properties [81]. Of particular importance for soft systems is gallium's ability to readily alloy with a wide variety of metals that can be utilized to reduce its melting point [78]. Two increasingly studied gallium alloys are Eutectic Gallium–Indium (EGaIn) and Galinstan, which are both liquids at room temperature. These liquid metals (LMs) maintain low toxicity and have near zero vapor pressure [78]. These alloys have been used for cooling of electronic components in replacement of thermal pastes and have been proposed for medical and pharmaceutical applications. EGaIn (T_m = 15.7 °C) is an alloy of gallium (75 wt %) and indium (25 wt %) with an electrical conductivity only an order of magnitude below copper (σ = 3.4 × 10^6 S m^{-1}) and a thermal conductivity of (k = 26.4 W m^{-1} K^{-1}) [82, 83]. Galinstan is composed of gallium, indium, and tin and can have various mixing ratios that can change its melting behavior with a typical composition of 68.5 wt % Ga, 21.5 wt % In, and 10 wt % Sn [82]. Galinstan often possesses a lower melting point (T_m = −19 to 10 °C), a similar electrical conductivity (σ = 3.5 × 10^6 S m^{-1}), and a relatively high thermal conductivity (k = 16.5 W m^{-1} K^{-1}) when compared to EGaIn but also possesses a lower boiling point (T_b ≈= 1300 °C) [82, 84].

EGaIn and Galinstan possess interesting mechanical properties that allow for a variety of use cases. Both EGaIn and Galinstan have low viscosities at about twice that of water (2.0 × 10^{-3} and 2.4 × 10^{-3} Pa s respectively) which allows for them to be processed similarly to many fluids [82]. However, in contrast to many traditional liquids, these LMs also rapidly form a solid oxide skin on their surface in the presence of oxygen, with a thickness on the order of nanometers [78, 85]. This oxide skin can provide a unique mechanical stability to the liquid that opens further avenues of processing. LM can be spray-coated onto surfaces utilizing similar methods to airbrushing with paints, making possible thin layers of the material to be patterned into conductive circuits [86]. The LMs can also be injected into channels that are patterned into a material utilizing methods such as applying positive pressure through a syringe or utilizing vacuum-assisted filling [87, 88]. There are also direct-printing techniques that to print 3d structures that have free-standing mechanical stability because of the oxide skin of the material [89]. While LMs can be processed independently, it has become increasingly common to employ these functional fluids as inclusions in elastomers for soft-electronic devices and robots, which allow for a combination of LM functionality while providing solid-state stability and containment for the LM.

2.2 Composite processing, structure, and properties

The combination of LM functionality and distinct mechanical response allows for unique processing and microstructure-dependent properties for LM–elastomer composites. One common method for the fabrication of LM composites is through shear mixing of liquid metals into an elastomer to produced microdispersed composite architectures [90–93]. During shear mixing, bulk LM breaks up into smaller liquid-metal droplets, where the final composite will be composed of discrete microdroplets within the elastomer matrix. The average diameter of LM droplet can be controlled by the mixing speed, with higher rates of mixing creating smaller mean droplets [53, 93, 94]. Through shear mixing alone, LM droplets are often in the range of 1 to several hundred μm, however other processing techniques can be used to further decrease particle size [53, 93, 94]. Nanoscale LM droplets can be fabricated prior to mixing into the elastomer by methods such as sonication which provide high-intensity energy for further droplet breakup [93–95].

Microdispersed LM composites can be created with exceptional functionality because of their ability to achieve high inclusion volume fractions of up to 90 % [96]. The LM in microdispersed samples exist as discrete droplet domains with insulating elastomer existing between each droplet leading to samples that can achieve enhanced thermal conductivities ($k \sim 1.5\,\mathrm{W\,m^{-1}\,K^{-1}}$) [59, 97]. However, these samples can remain electrically insulating (relative permittivity, $\epsilon_r \sim 1\text{–}100$) due to a lack of continuous electrical pathway which is particularly useful for heat transfer in electronics devices [59, 93].

The thermal conductivity and dielectric constant of these materials can be increased by elongation of the inclusions to create anisotropic samples, with high aspect ratio inclusions being responsible for some of the highest thermal conductivities reported in the literature for soft electrically insulating systems ($k \sim 13\,\mathrm{W\,m^{-1}\,K^{-1}}$) [52]. Utilizing LMs makes several unique opportunities to control inclusion shape. For example, the shape of microdispersed LM inclusions can be tuned through mechanical deformation (Figure 2a) where the droplets elongate in the direction of strain. [59] This allows for strain-dependent heat transfer characteristics that can be enabling for tunable electronics. However, in traditional thermoset elastomer composites the material will return approximately to its original state upon the release of the stress. One method for producing these highly elongated LM inclusions in a stress-free state is through the use of thermoplastic elastomers. These materials can be processed to contain highly elongated droplets in a stress-free state by deforming the sample to produced elongated inclusions, followed by annealing the thermoplastic elastomer to eliminate residual stress [52]. Additionally, the shearing of the uncured LM-polymer emulsion can elongate the droplets, while the rapid formation of gallium oxide imparts mechanical stability to the individual droplets which is believed to inhibit the ability of surface tension to rapidly return the droplet to its spherical shape. Upon curing the elastomer the LM droplets are fixed into shape. This method is especially useful since it can be used with additive manufacturing methods such as through direct ink-write printing (Figure 2b) which makes possible spatially programmable thermal properties by controlling print parameters [98, 99].

Although as-cast LM composites are typically electrically insulating, these materials can be made electrically conducting through post-processing. This type of electrical

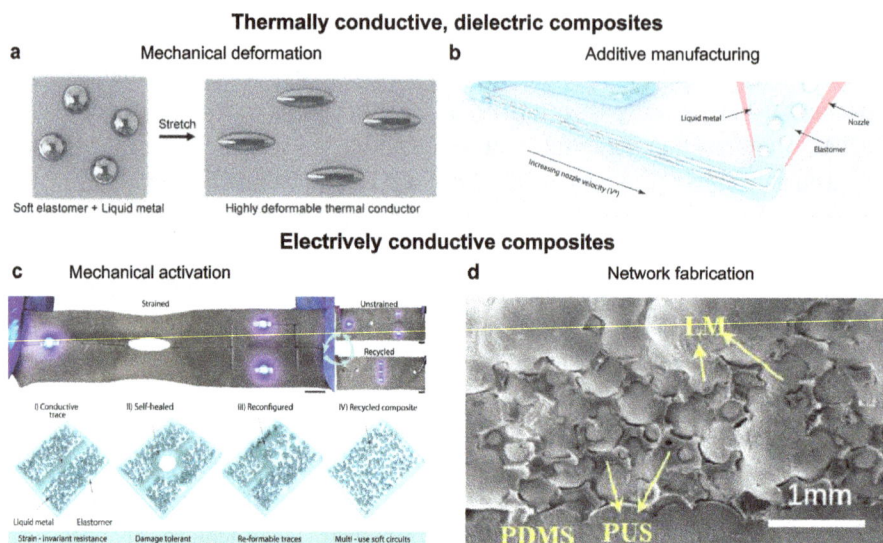

Figure 2: *Liquid-metal–elastomer composites.* Thermal conductivity of LM composites can be improved through microstructure control: *a)* Change in droplet shape by mechanical deformation. Open access article [59]. *b)* Programmed properties through additive manufacturing approach. Open access article [98]. Electrical conductivity can be enabled by formation of a connected LM network: *c)* Mechanical activation can cause a percolated network of LM inclusions. Open access article [100]. *d)* Fabrication techniques can be employed to create as-prepared connected networks. Reprinted from [101], with permission from Elsevier.

activation process was initially accomplished with LM droplets on a surface (not embedded in a composite) through a method referred to as mechanical sintering, where pressure is applied to cause rupture of the LM droplets that form a percolated conductive network [102–104]. LM composites can also be mechanically activated [9]. The activation pressure is dependent on the stiffness of the elastomer, the volume fraction of LM [53], and the size of the dispersed droplets [102]. Softer LM composites have generally been shown to be more challenging to activate because they resist rupture of the LM inclusions [105], whereas low-volume fractions can be challenging to activate and decreased droplet size increase the activation pressure required. LM composites can also be created to activate upon the application of strain, realizing 10^8-fold increases in conductivity after activation strains of 50 % [103]. Techniques such as embossing that utilize stamps to selectively apply pressure to the composite can be employed to create circuits with local conductive patterns by activation pressure (Figure 2c) [100]. Similarly, local activation can be achieved through writing techniques that apply sufficient surface pressure to hand-sinter [106]. Interestingly, these materials can also be self-healing, with damage events causing further percolation of droplets allowing for instantaneous healing of conductivity [8, 9, 100].

LM composites can also be fabricated as electrically conductive materials through fabrication techniques that create a continuous LM network. One method to accomplish this is to utilize porous materials such as sugar which can intake LM or elastomer into their pores under a vacuum. If LM is used to infiltrate the pores of the sugar, a subsequent step can be used where the LM is solidified below its melting temperature to hold shape and the sugar can then be dissolved in water [107]. The LM can then be penetrated and encapsulated with elastomer. If elastomer is used to infiltrate the pores of the sugar, the elastomer can be cured to hold shape, and LM can then be pulled into the elastomer [108]. Polymeric materials with 3D interconnected networks such as a polyurethane sponge (PUS) can also be purchased and infiltrated with LMs through vacuum to form a continuous network (Figure 2d) [101]. Additionally, multiphase systems can create electrically conducive composites as fabricated. This includes approaches where Ag flakes and LM are used together; the Ag flakes bridge the LM droplets and are electrically conductive [109, 110].

2.3 Liquid-metal–elastomer composite applications

The functional properties of LM composites are highly beneficial for soft electronic and robotic applications [14, 111, 112]. Thermally conductive LM–elastomer composites are capable of more efficient heat transfer than traditional elastomers, which can be utilized to improve the efficiency of soft machines [50, 52, 92, 113–115]. These materials are utilized to keep electronics such as high-powered LEDs cool and prevent overheating that can cause failure of the electronics [52, 97]. These materials have also been utilized to transfer heat away from heat-activated artificial muscles, keeping them cooler to enhance comfort for body-integrated devices as well as to speed up actuation cycles [52, 59]. Electrically conductive LM composites can be used to create mechanically compliant circuits capable of being interfaced with the human body [1, 116–118]. In one example, conductive composites were utilized to create a body-worn triboelectric nanogenerator, in which EGaIn was allowed to preferentially settle onto one side of the composite to create a conductive region and a dielectric region [119]. The triboelectric generator uses the EGaIn-rich side of the composite and human skin as electrodes separated by the layer of dielectric silicone elastomer. Deformation by the human body is then used to generate current capable of powering wearable devices [119]. LM can also be utilized for shape change and actuation by taking advantage of its solidification at low temperatures [120]. In this work, anisotropic composites that utilize an EGaIn-rich phase and an elastomer-rich phase can be utilized to create shape-morphing materials. The material can be prestretched and then frozen to solidify the LM, which causes the material to fix into place. Phase change in the LM can be caused by either heating or utilizing a laser on the composite, which causes the composite to change shape [120]. Electrically conductive soft composites can also be used to create self healing robots and electronics [8, 100, 121, 122]; in one cae, a soft robot was created by utilizing the LM composite as conductive wiring and integrating the robot with soft actuators and onboard control [9]. When damage events to the soft wiring occur, percolation of the LM inclusions follow, instantaneously repairing the conductive pathway without external intervention and allowing the robot to continue functioning [9].

3 Low melting-point alloys

Low melting-point alloys (LMPAs) are similar materials to liquid metals and can be used to impart conductivity to soft composites, as well as to allow for adaptive responses such as changes in stiffness and shape through phase change. LMPAs are generally considered to be metals with melting points below 300 °C [41], however in this work we will distinguish LMPAs from LMs by focusing on alloys with melting points below 100 °C that are solids above room temperature (25 °C). As such, we will not discuss LM alloys that are a liquid at room temperature such as EGaIn and Galinstan.

Gallium is a solid at room temperature [123], but its melting point of 29.8 °C is often too low for robotic applications since the material may experience an undesirable transition due to the heat from electronics components or weather. Bismuth-based LMPAs are strong options for soft robotic systems because there are several candidates with melting points between 60 and 150 °C [123]. A preferred LMPA used for stiffness tuning of materials is known as Field's metal which is an alloy of 32.5 % bismuth, 51 % indium, and 16.5 % tin by weight [91, 124, 125]. Field's metal is a desirable choice since it achieves a melting temperature of 62.5 °C [124], which is low enough that a transition can be accomplished at relatively small timescales with onboard heaters while high enough to withstand environmental temperatures without unplanned phase transition.

Much like LMs, LMPAs can possess desirable functional properties with Field's metal producing high thermal ($k \sim 19.2\,\mathrm{W\,m^{-1}\,K^{-1}}$) and electrical ($\sigma = 1.92 \times 10^6\,\mathrm{S\,m^{-1}}$) conductivities at room temperature [123]. The high thermal conductivity and phase-change behavior of Field's metal has made it an attractive candidate for heat transfer in electronics and larger systems. LMPAs can also be utilized as thermal buffers for electronics that can absorb or release heat during phase change [126]. The low melting point and high electrical conductivity has also made Field's metal a desirable candidate for non-toxic, lead-free solder [127]. Similarly, Field's metal has been studied for its ability to be used as electrical contacts through 3D printing methods. While electrodes on traditional electronics such as solar cells are often deposited through thermal evaporation, Field's metal has been shown to be printable through direct ink-write printing to produce electronics with similar performance to the state-of-the-art counterparts [128]. The emergence of soft electronics and robots has created new opportunities for LMPAs that have enabling properties for these systems, resulting in the use of LMPAs for soft composites. Solid-state LMPAs are particularly enabling for soft robots since they can be transitioned from the solid to the liquid state with the input of thermal energy. By utilizing these materials as inclusions in elastomer composites, the solid–liquid phase transition can be leveraged to tune stiffness and create reconfigurable materials that can be locked into multiple shapes [123, 124, 129–133].

3.1 Fabrication and structure of LMPA composites

The integration of LMPAs into elastomers allows for an increase in thermal and electrical conductivity of the composite as well as large magnitude stiffness change, where the elastomer contains the LMPA during its liquid state. The fabrication strategies of LMPAs are diverse, owing to its ability to be processed in either their solid or liquid form. One method for creating these composites is through encapsulation of the LMPA, where the LMPA is solidified and elastomer is placed around the solid-state LMPA [129]. The LMPA can be fabricated into complex shapes by molding of the LMPA in its liquid state, followed by subsequent cooling. Another method of fabricating these composites is by introducing the LMPA into prepatterned elastomer channels (Figure 3a) which can be accomplished through several methods [134]. One method is by patterning an elastomer with channels that contain both inlet and outlet holes [135, 136]. The LMPA can be melted over the inlet chamber and positive pressure can force the LMPA through the channel which can then be cooled and sealed [54]. Another method is to create channels with a single access port [60]. LMPA can be placed on the access port and melted in a vacuum oven, which will pull air from the channels that are then filled with LMPA. Another method is to utilize a heated syringe. By outfitting a resistive heater onto a syringe, the LMPA can be melted and injected into channels in its liquid state.

Recently, composites have also been made by utilizing LMPA microparticles (Figure 3b) that are dispersed throughout a polymer matrix. LMPAs can be processed into particles through several methods including shear mixing and sonication. One method utilizes a combination of an overhead mixer and homogenizer. Field's metal is added to a container with heated water kept at 90 °C in an oil bath. The homogenizer is utilized to shear the bulk material into smaller particles while the overhead mixer is utilized to keep particles suspended and will mixed to ensure passage through the homogenizer. The particles are allowed to cool during overhead mixing. This method produced particles with a large size distribution from 10 to 800 μm [124]. Field's metal microparticles have been also been fabricated by a technique known as shearing liquids into complex particles (SLICE) [140]. Field's metal was added to a 5 % aqueous acetic acid solution and heated to 95 °C before being shear mixed to produce particles with average diameters of 12.6 ± 8.9 μm. Another way of producing Field's-metal particles is through sonication. Fields metal can be added to water at 70 °C, and an ultrasonicator can be used to break up the droplets. This has been utilized to produce Field's metal droplets with average diameter of 1.19 μm [141]. A similar method utilizing a large viscosity continuous phase (polyalphaolefin) instead of water has been used to create fields metal nanoparticles

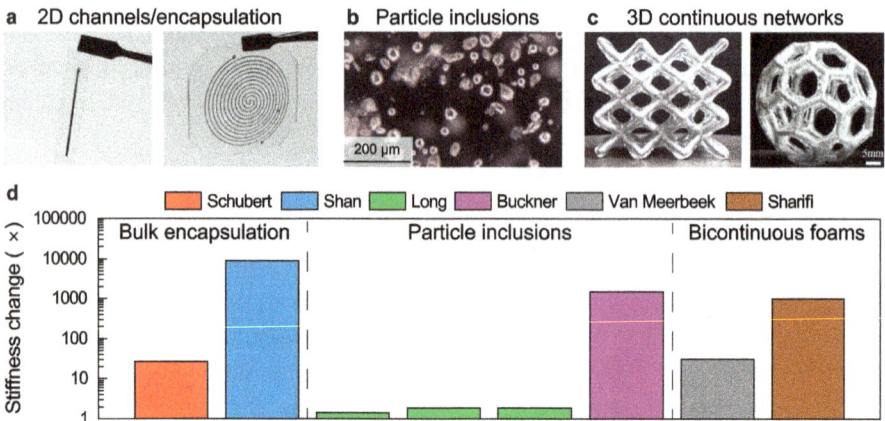

Figure 3: *Low melting-point alloy composites.* Composites have been developed through: *a)* Fabrication of LMPA channels in elastomer. Reprinted (adapted) with permission from [134]. Copyright 2021 American Chemical Society. *b)* The creation of LMPA particles for dispersion into elastomer. *c)* 3D continuous networks of LMPA in elastomer. Reprinted from [137], with permission from Elsevier. *d)* Stiffness change of LMPA composites of various microstructures, data from various sources [54, 55, 124, 129, 138, 139].

with average diameters as low as 15 nm [142]. After creating the LMPA particles, they can then be dispersed through the elastomer by mixing, where the composite can then be set into a desired shape through molding [124]. In addition to molding, the shape of these LMPA microparticle composites can be created through additive manufacturing techniques such as direct ink-write printing [138].

LMPA-elastomer composites have also been created in unique structures such as 3D continuous networks (Figure 3c) [137]. These can be created through a multistep process where first a sacrificial lattice structure is 3D printed and an investment mold is cast around the part. Next, the sacrificial lattice can be burnt out of the mold, and the mold can be infiltrated with LMPA at elevated temperature. Lastly, the investment mold can be removed from the LMPA, and the LMPA can be coated with elastomer [137]. In addition to macroscale geometries, 3D continuous networks can also be formed at the microscale in the form of bi-continuous foams [55, 139]. Making these foams is a multistep process, where open-cell foams can be created by mixing uncured elastomer precursor with grains of salt. The elastomer can then be immersed in water to dissolve the salt inclusions and leave pores [139]. The LMPA is then heated into its liquid state, and the elastomer foam is submerged in the LMPA under a vacuum to draw the LMPA into the material. By controlling pore size, the mechanical properties of the composite can be tuned, where larger porosity can result in a greater dependence on the modulus of the LMPA and a lower dependence on the modulus of the elastomer. The opposite processing technique can also be used, where salt can be added to the LMPA to produce an LMPA foam which can then be infiltrated with liquid elastomer precursor prior to curing [55]. This technique produces a more continuous LMPA structure.

3.2 LMPA-composite stiffness tuning

LMPA composites possess some of the highest magnitudes of stiffness tuning among active materials due to their ability to tune from liquid-state inclusions with low storage modulus to solid-state inclusions on the order of GPa [74, 123, 133]. The response time of LMPA composites is generally dependent on the melting point of the LMPA used, the mass of LMPA in the composite, and the heating mechanism utilized to drive the phase transformation, and it has been reported to be as fast as 0.5 s or as slow as ~10 min [54, 60]. The response time to cool the composite from the liquid to the solid state can also be on the order of minutes. The stiffness tuning capability of LMPA composites can be up to 10,0000× for encapsulated systems (Figure 3d) [54, 55, 124, 129, 138, 139]. Dispersed-particle composites and other forms of LMPA structure have also shown large changes in stiffness of greater than 1000× [55]. One additional approach has been to create soft composites from undercooled Field's metal particles, which are metastable at room temperature. This traps the particles in a liquid state at room temperature, and upon mechanical perturbation the particles rapidly solidify which results in a 300 % increase in Young's modulus [143].

3.3 LMPAs and soft robots

LMPA composites are particularly enabling for soft robots that must undergo changes in shape for multifunctional use cases or require stiffening for changes in mechanical stability. Variable stiffness fibers have been fabricated that utilize LMPA as a core that is encapsulated with elastomer and heating elements to enable shape change in robotic systems [62, 144]. These fibers can also show a shape-memory effect where the rigid LMPA allows for shape retention, while elastic restoration by the polymer can allow for shape reversibility upon melting of the LMPA [144]. These fibers can be used as joints in shape-morphing robotic systems that change shape to aenable multiple modes of transport, such as flying or driving (Figure 4a) [62]. Low melting-point alloys can also be utilized to create soft robotic manipulators [145]. In this case, a syringe pump is attached to an elastomer manipulator, with both the elastomer and syringe wrapped in resistive heating wires. The actuation is pressure driven by utilizing the LMPA in its liquid state, and the deflection of the manipulator can be controlled by the amount of liquid LMPA supplied from the syringe. The LMPA can also be solidified to increase its stiffness make possible higher forces of deflection. LMPA–elastomer composites can also be utilized to affect adhesion in order to create soft robotic grippers, where the gripping force can be reversibly controlled through stiffness tuning [146, 147]. LMPA bicontinuous foams can also be used to create soft actuators capable of both high degrees of bending and shape-locking effects [139]. By integrating the LMPA foam with a pneumatically driven elastomer, the actuator can change shape and rigidity. The LMPA foam can decrease its stiffness by heating that allows the pneumatic actuator to change shape with pressure,

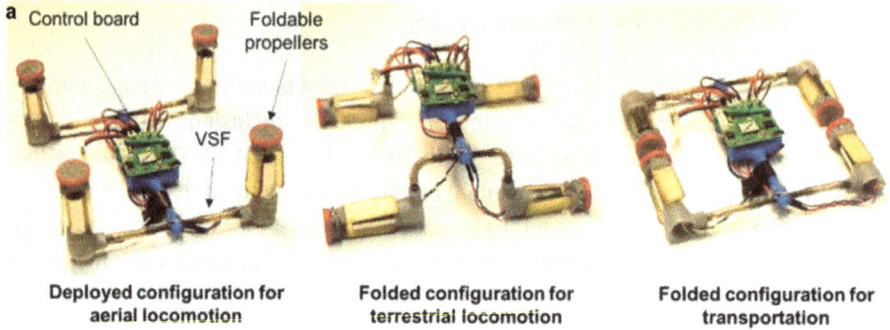

a

Control board | Foldable propellers

VSF

Deployed configuration for aerial locomotion

Folded configuration for terrestrial locomotion

Folded configuration for transportation

Figure 4: *Soft robot LMPA applications. a)* LMPAs for shape-changing multifunctional robots. Open access article [62].

and the actuator can then solidify into shape by allowing the LMPA foam to cool, which allows for a greater ability to manipulate objects [139]. By utilizing LMPA composites to form the main body of robots, large-scale shape change can be accomplished that enables deployable machines and multimodal robots [60]. In addition to stiffness tuning effects, stretchable electronics have been achieved 3D-printing using Field's metal onto soft substrates, where it can be deposited into patterns that allow for stretchability in its solid state [148].

4 Magnetic elastomer composites

Magnetic materials play an important role in traditional electronic devices and robots for functions like power transfer and actuation. To create mechanically soft magnetic materials for soft systems, magnetic fluids or magnetic particles in elastomers are often utilized. Magnetorheological fluids (MRFs) are a class of adaptive fluid created by dispersing magnetic powders within a nonmagnetic carrier fluid; they have been studied for their field-dependent rheology such as tunable viscosity and yield stress for their use in variable damping and energy-dissipation systems [149–153]. The tunable properties of MRFs are enabled by the magnetic-particle response to a magnetic field. When a magnetic field is applied, the particles rotate within the carrier fluid and largely align in the direction of the field, forming a network structure of particles. This network structure resists deformation from an applied load, and the mechanical response portrays an initial linear region of stress and strain up until the yield stress of the MRF [149, 150]. After reaching the yield stress, the magnetic particles begin to slip past each other and a flow of the MRF is observed, although there is still an increase in the resistance to flow, and therefore viscosity, when compared to a nonactivated sample due to the continued particle interaction with the applied field [149, 154, 155]. MRFs are particularly enabling for machine systems not only because of their tunable rheology, but also because of their rapid response time (~10–100 ms) [150, 151].

Magnetic elastomers or magnetorheological elastomers (MREs) are a class of adaptive solids created by dispersing magnetic powders within a nonmagnetic elastomer. Magnetic elastomers can introduce both functional characteristics and an adaptive mechanical response. For magnetic functionality, the addition of magnetic material to an elastomer provides increased magnetic permeability which allows for control over the magnetic field's magnitude and direction, which is important for the development of magnetic circuits [42, 156]. Magnetic elastomers also possess unique magneto-mechanical properties such as magnetostriction: the change in length or volume in response to a magnetic field and a magnetorheological effect that leads to tunable mechanical properties in response to a magnetic field [157–160]. MREs have been similarly studied for their use as adaptive materials for applications such as actuators and automotive components capable of variable damping properties through their magnetorheological response [161–164]. These composite materials are the solid analog to MRFs and have traditionally been made as rigid particle composites, with micron-scale magnetic powders dispersed throughout an elastomer matrix [165]. MREs overcome many limitations observed in MRFs because the elastomer provides structural stability to the particles, which allows for material usage without the need for a containment vessel, and eliminates the time-dependent settling of particles than can be seen in their fluid counterpart. When a magnetic field is applied to these materials, they rapidly undergo changes to their mechanical response such as through storage and loss modulus, making them ideal for tunable stiffness and energy-dissipation materials [166, 167]. The mechanism behind this tunability is similar to that seen in MRFs with some key differences. Upon an application of a magnetic field, the embedded particles attempt to align in the direction of the field, however, in contrast to MRFs, the particles are embedded in a solid-state matrix, which limits their ability to fully align. However, the magnetic response of these particles provides an adaptive material response in the form of magnetostriction [157], as well as a change in stiffness because of the resistance to deformation of the embedded inclusions [168]. These materials have been shown to have some of the fastest response times (<10 ms) among adaptive materials, allowing for rapid response and tuning of their mechanical properties [169, 170]. In this section, we will discuss the fabrication and microstructure of MREs, as well as the implications towards magnetic functional response and adaptive mechanical response. The applications of these materials towards soft electronics and robots will then be detailed.

4.1 MRE structure

To allow for a controllable and useful response, MREs should possess a higher magnetic permeability to allow for a large magnetic response, a high saturation magnetization that allows for the material to adapt its response over a large range of applied fields, and a low remnant magnetization that governs the material's ability to rapidly reverse its response in the absence of a magnetic field [171]. For these reasons, iron is often chosen as the magnetic filler due to its high permeability, maximum saturation magnetization among elements, and low remnant magnetization, as well as its relative cost when compared to other engineered magnetic powders [171, 172].

MREs are often measured by their change in modulus under an applied field, called the MR effect, which can either be measured as an absolute (ΔE_c) or relative (i. e., $\frac{\Delta E_c}{E_{c,B=0}} \times 100\,\%$) value, where $E_{c,B=0}$ is the composite modulus in the absence of a magnetic field. In order to improve the MR effect, MREs can be optimized through the material structure, which is often done through magnetic particle size [173–176], shape [157, 167, 177], volume fraction [160, 178–180], and the microstructure distribution throughout the MRE [56, 159]. The particle size for most MRE systems is on the order of microns, however recent work has shown MR effects utilizing nanoparticles as well [176]. The volume fraction of magnetic particles has a significant effect on the magnetic response of MREs, where higher volume fractions of particles lead to higher relative permeability and higher activated modulus [156, 160]. However, larger particle volume fractions also increase the zero-field modulus of the MRE, which can lead to a lower relative MR effect [160].

Rigid particle composite (RC) MREs are often made by evenly distributing the particles throughout the elastomer [181], however one way to improve MRE properties is through the development of anisotropic composites (Figure 5a). Anisotropic MREs can be made by curing the sample in the presence of a magnetic field, which preferentially aligns the particles prior to curing [171]. This results in samples with higher relative permeability in the direction of the field and a larger magnetorheological effect [182, 183]. The degree of anisotropy is larger in lower-loaded particle systems, where lower volume fractions show significant anisotropy and differences when compared to isotropic systems. However, at high volume (>30 %) fractions, the particles will form a more complex network structure similar to those seen in isotropic MREs or activated MRFs [160, 184].

In addition to magnetic particles, magnetorheological fluids have recently been used as inclusions in elastomers to increase their MR effect. These magnetorheological fluid composite (FC) MREs have demonstrated some of the largest relative MR effects seen in MRE systems [57, 58, 186, 187]. FC MREs have been made through several methods including bulk encapsulation in an elastomer [186], additive manufacturing approaches to selectively place shape-controlled millimeter scale fluid domains into desired positions [58], and recently microdispersed MRFs have been added to elastomers (Figure 5b) [56, 57, 63, 187]. For bulk encapsulation, the MRF is filled into an elastomer mold, which is than sealed with elastomer. The elastomer provides a structural component to the MRE, while the mechanical response is tuned with the bulk the fluid phase. This material showed higher tunability in mechanical response when compared to traditional RC MREs [186]. To provide further microstructural control, additive manufacturing approaches have also been used to create FC MREs, where mm-scale fluid

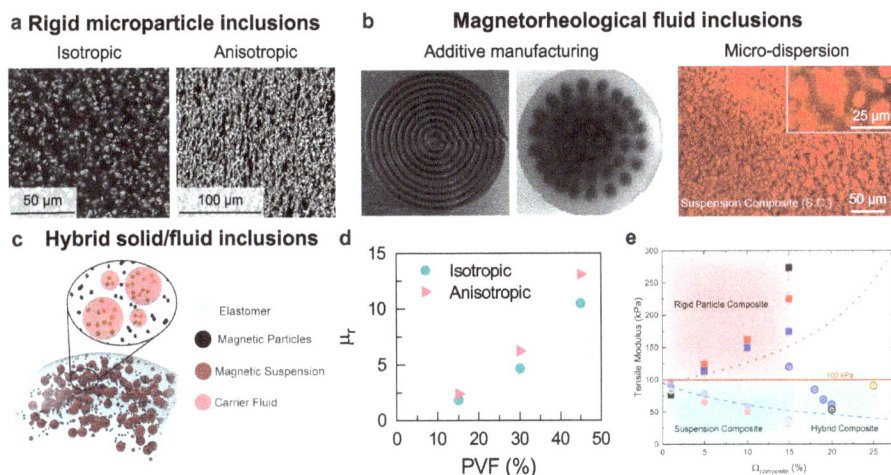

Figure 5: *Magnetorheological elastomers for functional and active materials. a)* Microstructure of rigid-particle composite (RC) MREs as either isotropic or aligned as anisotropic with a magnetic field. Left image adapted from an open access article [63]. *b)* Magnetorheological fluid composite (FC) MREs can be shape-programmed by additive manufacturing approaches. Reprinted from [58], with permission from Elsevier. They can also be created by microdispersing magnetorheological fluid droplets though mixing. Reprinted (adapted) with permission from [56]. Copyright 2020 American Chemical Society. *c)* Solid/fluid hybrid composite (HC) MREs utilize a combination of particles and fluids. Reprinted (adapted) with permission. Copyright 2020 American Chemical Society [56]. *d)* Permeability of isotropic and anisotropic MREs using data from Ref. [185]. *e)* Effect of magnetic particles and fluids on MRE zero-field modulus. Reprinted (adapted) with permission from [56]. Copyright 2020 American Chemical Society.

domains can be selectively added to desired positions within an elastomer matrix. By utilizing 3D printing, the shape and position of MRF domains can be prescribed, allowing for spatial tuning of magnetorheological properties, as well as adding anisotropy and complex domain shapes to MREs. The control over size and shape were shown to have a strong influence on the adaptive properties of the MRE [58]. In addition to these approaches, microdispersed FC MRE composites have recently been produced. These materials are made by mixing MRF into uncured polymer prior to curing and have shown some of the highest changes in stiffness of MRE materials (up to 30x) [56, 57, 63]. Additionally, the microstructure of these materials was observed under X-ray tomography, which showed alignment of individual MRF domains in response to a field, providing insight into the response of MRF inclusions. Other advances have been the addition of both solid and fluid magnetic fillers to elastomers which have resulted in hybrid composite MREs (Figure 5c) with stiffness changes of up to 50x their zero-field modulus [56, 63].

4.2 MRE properties

The mechanical and functional properties of MREs are influenced by the material structure. For isotropic composites with spherical inclusions, the effective relative permeability (μ_{eff}) of magnetic-elastomer composites is effected by the magnetic permeability of the magnetic inclusions (μ_i), the magnetic permeability of the elastomer (μ_e), and the volume fraction (f) of magnetic particles [188]. The effect of relative permeability can be estimated by models such as the Maxwell Garnett equation (Equation (1)) which assumes evenly distributed particles with only weak interaction between particles [189]. This model works well for composites of lower volume fractions, but begins to underpredict the permeability as the volume fraction increases and more interparticle interactions are observed [185].

$$\mu_{\text{eff}} = \mu_e + 2\mu_e \frac{\mu_i - \mu_e}{\mu_i + \mu_e - f(\mu_i - mu_e)}. \tag{1}$$

As previously discussed, the magnetic permeability of composites can be increased by utilizing a magnetic field to impart anisotropic alignment of particles through the elastomer, especially if anisotropic particle geometries are incorporated. Preferential alignment of magnetic particles make possible magnetic shorting of the elastomer matrix between particles resulting in higher magnetic-field strength arising from higher effective permeability [156]. Figure 5d demonstrates the difference in relative permeability as volume fraction is increased for both isotropic and anisotropic composites [185]. In this case, the relative permeability of the composite increases by 25 % when the magnetic particles are aligned with a magnetic field. For isotropic MREs, the relative permeability has not been shown to be dependent on whether the composite is made with rigid particles, magnetic fluids, or a combination of the two, and is instead dependent on the total volume fraction of magnetic content in the material [56].

The zero-field mechanical properties of MREs are also heavily dependent on the volume fraction of magnetic inclusions as well as the type of magnetic material used (Figure 5e) [56]. As previously discussed, MREs have traditionally been made as rigid particle composites, and recently as magnetorheological fluid composites as well as hybrid composite material architectures. The composite modulus (E_c) of MREs can generally be predicted by an Eshelby model (Equation (2)), where E_i is the modulus of the magnetic inclusions, E_e is the modulus of the elastomer, and ϕ is volume fraction of inclusions [190].

$$E_c = E_e \frac{1 + \frac{2E_i}{3E_e}}{\left(\frac{2}{3} - \frac{5\phi}{3}\right)\frac{E_i}{E_e} + \left(1 + \frac{5\phi}{3}\right)}. \tag{2}$$

For rigid particle composites, ϕ is equal to the volume fraction of magnetic particles ($\Omega_{\text{composite}}$), and E_i is equal to the elastic modulus of the particles. For magnetorheological fluid composites, ϕ is equal to Ψ, the volume fraction of magnetorheological fluid.

For FC MREs, E_i has been shown to be equal to zero, indicating that the magnetic particles in the MRF can remain encapsulated in the dispersed microdroplets and decrease the modulus of the composite [56]. For RC MREs, the composite modulus increases as the volume fraction of solid increases. Different powder geometries can also have an effect on the composite modulus, such as flake-like geometries that can further increase the stiffness of the composite at lower volume fractions than spherical powders [42, 56].

4.3 MREs and soft robots

The rapid magnetorheological and shape change response of MREs is enabling for soft electronics and adaptive machines [73, 191]. For example, the shape translation through magnetic interaction can be used for soft actuation and movement [161, 192–194] and for the development of reconfigurable electronics and robots [195, 196]. MREs have been used for haptic feedback systems that can be spatially actuated with a magnetic field to produce indentation and stiffness change at desired locations (Figure 6a) [176]. In this case, the material was made as an Arduino-controlled interface capable of deforming six buttons at different positions on the surface of the device, where the buttons were activated with a magnet to spell letters in braille. Reconfigurable electronics can be made by utilizing MREs with a conductive pathway that acts as a switch when magnetically actuated (Figure 6b) [196]. The MRE is utilized with a conductive composite created with silver-coated nickel microparticles on one side. Upon activating with a magnet, the material can actuate towards an LED circuit to complete the conductive pathway and switch the LED on. The stiffening material response of MREs can also be utilized to rapidly

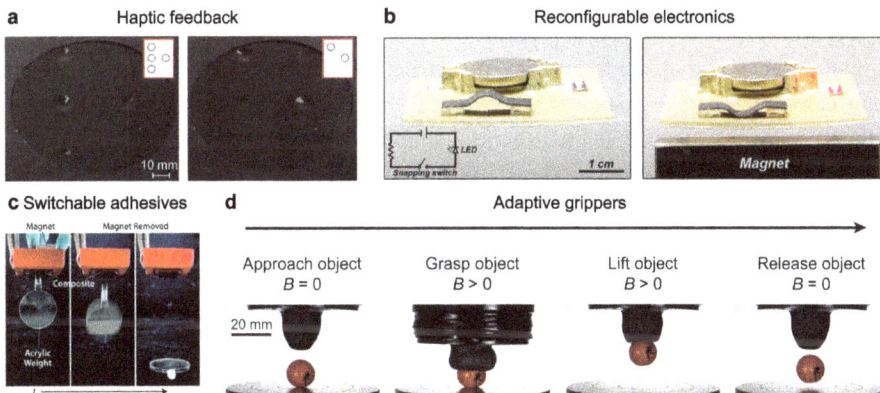

Figure 6: *MRE soft-robot applications. a)* MREs can rapidly respond to a field for machine actuation such as in haptic feedback systems. Reprinted (adapted) with permission from [176]. Copyright 2022 American Chemical Society. *b)* Shape change can be used for reconfigurable electronic devices. Open access article [196]. *c)* The change in stiffness of MREs allows for rapidly reversible adhesion. Open access article [187]. *d)* Stiffness tuning can also be used to create adaptive robotic grippers. Open access article [63].

tune adhesion for switchable adhesives [197–199], where, in many cases, the on state of the adhesive is the magnetically actuated state and the off state realizes a decrease in adhesion and stiffness (Figure 6c) [187]. Magnetic elastomers can also be used for soft robotic grippers that can adapt to the shape of fragile objects and then firmly grasp the objects upon the application of a magnetic field [63, 200, 201]. In Figure 6d a magnetic elastomer gripper has been integrated with an electromagnet to create a gripper that can passively conform to a variety of object geometries while soft, but can then rapidly (~20 ms) increase its stiffness by up to 50x to grasp the object upon application of a magnetic field [63]. Additionally, recent research has used hard magnetic powders as inclusions that can be utilized for greater control over material motion which can be used for locomotion and actuation of miniature soft robots [202–204]. A miniature robot has been created using hard magnetic powders that achieves multiple modes of motion, allowing it to crawl, walk, roll, swim, and even grasp objects depending on the direction and strength of the field [202].

5 Multicomponent soft composites

Multifunctional composites enable notable advances for soft robotic applications including soft wiring, heat transfer, and adaptive shape and mechanical responses. While the advancement of each individual property is important, soft robots often utilize combinations of functional and active properties which can be enhanced through the creation of multicomponent material systems.

5.1 Multicomponent composite architectures

Liquid metal can be utilized to create soft wiring for electronic components, however some electronics can be enhanced with the addition of integrated functional materials. Inductive power transmitters feature a coil-shaped electrical pathway that transmits a magnetic field to an inductive receiver. Traditional wireless power devices feature a copper coil as well as backplane made of magnetic material, which increases the efficiency of power transfer. Stretchable inductors can be made by utilizing a multicomponent system, which utilizes a liquid-metal inductor coil integrated with a high permeability magnetic elastomer backplane. (Figure 7a) [42]. Multicomponent materials can also be used to tune material response, such as when magnetic particles are used in addition to liquid metals. An anisotropic liquid-metal magnetorheological elastomer was developed that utilizes magnetic alignment to enhance the properties of the liquid-metal composite. Alignment both significantly increases the undeformed electrical conductivity of the composite in all directions as well as introduces both positive and negative piezoconductivity where the conductivity can increase or decrease with applied strain depending on the direction of the electrical measurement (Figure 7b) [205].

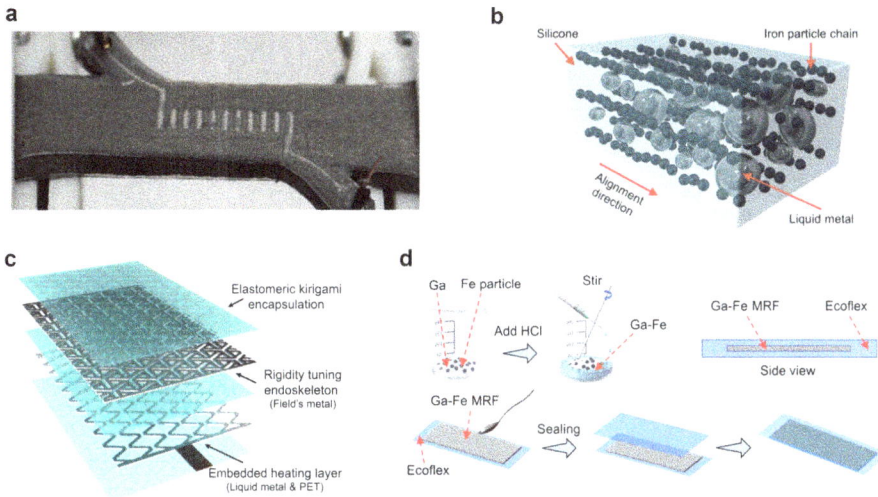

Figure 7: *Multicomponent soft composites. a)* Liquid-metal conductive pathway integrated with magnetic elastomer composite to increase inductor efficiency. Reprinted (adapted) with permission. Copyright 2015 American Chemical Society [42]. *b)* Magnetic particles integrated with liquid-metal droplets for tunable functionality. Reprinted from [205], with permission from Elsevier. *c)* Low melting-point alloy integrated with liquid metal-based resistive heater. From [60]. Reprinted with permission from AAAS. *d)* Magnetorheological fluid utilizing LMPA as a carrier fluid capable of rapid magnetorheological response and further stiffness tuning through phase change. Open access article [206].

Another multicomponent elastomer strategy is to utilize both liquid metals and low melting-point alloys. Continuous liquid-metal channels can be utilized as resistive heaters for low melting-point alloys, allowing for a compliant onboard heater capable of large deformations. By combining LMs and LMPAs in a multilayered system, shape and stiffness tuning materials can be created that allow for reconfigurability in soft robots. (Figure 7c) [60].

For stiffness-tuning materials, MREs show some of the fastest response times among stiffness-tuning materials, however other material systems such as LMPAs often exhobit a larger magnitude response. One material system utilizes liquid-state LMPA as the carrier fluid for a magnetorheological fluid that can also change phase to become a solid [206]. By utilizing a combination of LMPA and magnetic particles, rapid response time can by achieved in stiffness-tuning materials while also allowing for higher magnitudes of change in elastic modulus by solidifying the LMPA (Figure 7d).

5.2 Multicomponent composites in robots and electronics

Multicomponent materials allow for unique combinations of properties for new forms of soft electronic devices and robots. For example, shape-memory alloys (SMAs) are an active material that can change between crystalline phases in response to heat. This can

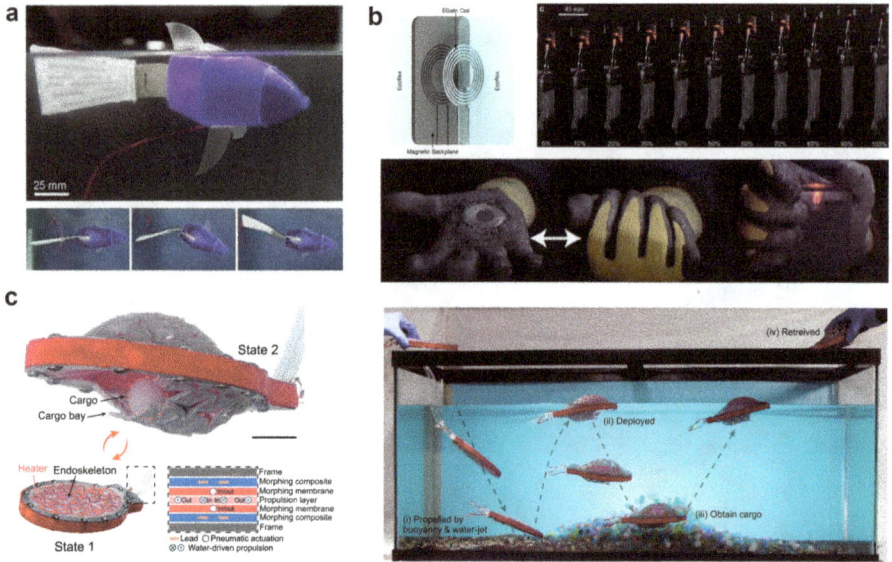

Figure 8: *Multicomponent material systems enable next-generation electroincs and machines. a)* Liquid-metal heat-transfer material enables robotic movement via an SMA actuator. Open access article [59]. *b)* Magnetic elastomer backplanes improve the power transfer efficiency of LM-based wireless power-transfer devices. Reprinted (adapted) with permission. Copyright 2020 American Chemical Society [56]. *c)* Liquid-metal heater allows for reversibility of shape-morphing material based on LMPA. From [60]. Reprinted with permission from AAAS.

give rise to an actuation effect, where the material will deform in response to heat and return to its original configuration in response to cooling. However, the ability to rapidly cycle SMA actuation requires a method to cool the SMA. Liquid-metal elastomers with high thermal conductivity have been utilized as heat dissipating layers for SMAs that are compliant enough for actuation while removing heat to improve actuation. Figure 8a shows the capability of high thermal conductivity LM composites to increase the functionality of an SMA actuator on a soft robotic fish [59]. Further, robotic fabrics can be created through LMPA filaments and SMA wires for load-bearing structures that morph into multiple configurations [207]. These types of robotic skins can also be applied to tensegrity robotics [208, 209].

Utilizing multiphase magnetic composites in addition to liquid-metal wiring allows for soft and conformal inductors capable of wireless power transfer. Liquid-metal coils can be added to elastomers to create inductors capable of wirelessly transmitting and receiving power. Magnetic elastomers that utilize iron particles as inclusions can act as a magnetic backplane, which improves the power-transfer efficiency of these devices. By incorporating liquid-metal coils into magnetic backplanes, wireless power transfer systems can be created. These all soft transmitted and receiver systems can transfer power up to 100 % strain; they can be utilized in high-deformation regions of the human body, such as the palm of a human hand. Figure 8b shows the ability of LM-magnetic elastomer inductors to be used for wearable wireless power transfer to rigid wireless receivers [56]. Additionally, magnetoactive phase change materials with magnetic particles embedded in liquid metal can reversibly switch between the solid and liquid states. The composite can also morph by changing shape, splitting, or merging by controlling the composite magnetically in the liquid state [210].

Liquid-metal and low melting-point alloy multilayer composites allow for rapid and reversible shape morphing in soft robot systems. These can be enabling for the creation of deployable machines, which can start in a flat and compact shape before transforming into a functional shape. These machines can interact with their environment such as through obtaining cargo, and reverse back into their compact shape after use. Figure 8c shows a deployable submersible machine diving underwater and obtaining cargo before being retrieved and reverting to its original shape [60].

6 Conclusions

Soft functional composites are essential components for emerging soft robots. Soft functional materials such as liquid-metal composites and magnetic elastomers allow for high thermal, electric, and magnetic properties in a compliant and deformable material. Additionally, the study of soft functional devices has lead to the development of adaptive materials that can change their properties and shape in response to engineering stimuli. Magnetorheological elastomers and low melting-point alloys can be utilized to allow for tunable stiffness and reconfigurability in soft machines. By combining multifunctional materials in multicomponent systems, properties from several forms of soft functional composite can be combined, which opens intriguing opportunities for material properties and soft robot applications, such as tunable material properties including conductivity and piezoresistivity, increased efficiency of wearable devices such as wireless chargers, and further degrees of stiffness tuning by combining LMPAs and magnetic materials.

Bibliography

[1]   Park, Y. G.; Lee, G. Y.; Jang, J.; Yun, S. M.; Kim, E.; Park, J. U. Liquid metal-based soft electronics for wearable healthcare. *Adv. Healthc. Mater.* **2021**, *10* (17), 2002280.
[2]   Li, H.; Wang, J.; Fang, Y. Bioinspired flexible electronics for seamless neural interfacing and chronic recording. *Nanoscale Adv.* **2020**, *2* (8), 3095–3102.
[3]   Whitesides, G. M. Soft robotics. *Angew. Chem., Int. Ed.* **2018**, *57* (16), 4258–4273.
[4]   Kim, S.; Laschi, C.; Trimmer, B. Soft robotics: a bioinspired evolution in robotics. *Trends Biotechnol.* **2013**, *31* (5), 287–294.

[5]   Laschi, C.; Mazzolai, B.; Cianchetti, M. Soft robotics: technologies and systems pushing the boundaries of robot abilities. *Sci. Robot.* **2016**, *1* (1), eaah3690.

[6]   Polygerinos, P.; Correll, N.; Morin, S. A.; et al. Soft robotics: Review of fluid-driven intrinsically soft devices; manufacturing, sensing, control, and applications in human-robot interaction. *Adv. Eng. Mater.* **2017**, *19* (12), 1700016.

[7]   Bao, G.; Fang, H.; Chen, L.; et al. Soft robotics: Academic insights and perspectives through bibliometric analysis. *Soft Robot.* **2018**, *5* (3), 229-41.

[8]   Bartlett, M. D.; Dickey, M. D.; Majidi, C. Self-healing materials for soft-matter machines and electronics. *NPG Asia Mater.* **2019**, *11* (1), 21.

[9]   Markvicka, E. J.; Bartlett, M. D.; Huang, X.; Majidi, C. An autonomously electrically self-healing liquid metal–elastomer composite for robust soft-matter robotics and electronics. *Nat. Mater.* **2018**, *17* (7), 618–624.

[10]   Kang, J.; Tok, J. B. H.; Bao, Z. Self-healing soft electronics. *Nat. Electron.* **2019**, *2* (4), 144–150.

[11]   Rogers, J. A.; Someya, T.; Huang, Y. Materials and mechanics for stretchable electronics. *Science* **2010**, *327* (5973), 1603–1607.

[12]   Xiloyannis, M.; Alicea, R.; Georgarakis, A. M.; et al. Soft robotic suits: State of the art, core technologies, and open challenges. *IEEE Trans. Robot.* **2021**, *38* (3), 1343–1362.

[13]   Hawkes, E. W.; Majidi, C.; Tolley, M. T. Hard questions for soft robotics. *Sci. Robot.* **2021**, *6* (53), eabg6049.

[14]   Dickey, M. D. Stretchable and soft electronics using liquid metals. *Adv. Mater.* **2017**, *29* (27), 1606425.

[15]   Bartlett, M. D. Liquid assets for soft electronics. *Nat. Mater.* **2021**, *20* (6), 714–715.

[16]   Kim, D. H.; Lu, N.; Ma, R.; et al. Epidermal electronics. *Science* **2011**, *333* (6044), 838–843.

[17]   Ma, Y.; Jang, K. I.; Wang, L.; et al. Design of strain-limiting substrate materials for stretchable and flexible electronics. *Adv. Funct. Mater.* **2016**, *26* (29), 5345–5351.

[18]   Baca, A. J.; Ahn, J. H.; Sun, Y.; et al. Semiconductor wires and ribbons for high-performance flexible electronics. *Angew. Chem., Int. Ed.* **2008**, *47* (30), 5524–5542.

[19]   Reyes-Martinez, M. A.; Crosby, A. J.; Briseno, A. L. Rubrene crystal field-effect mobility modulation via conducting channel wrinkling. *Nat. Commun.* **2015**, *6* (1), 6948.

[20]   Kim, D. H.; Ahn, J. H.; Choi, W. M.; et al. Stretchable and foldable silicon integrated circuits. *Science* **2008**, *320* (5875), 507–511.

[21]   Khang, D. Y.; Jiang, H.; Huang, Y.; Rogers, J. A. A stretchable form of single-crystal silicon for high-performance electronics on rubber substrates. *Science* **2006**, *311* (5758), 208–212.

[22]   Bowden, N.; Brittain, S.; Evans, A. G.; Hutchinson, J. W.; Whitesides, G. M. Spontaneous formation of ordered structures in thin films of metals supported on an elastomeric polymer. *Nature* **1998**, *393* (6681), 146–149.

[23]   Choi, Y. S.; Hsueh, Y. Y.; Koo, J.; et al. Stretchable, dynamic covalent polymers for soft, long-lived bioresorbable electronic stimulators designed to facilitate neuromuscular regeneration. *Nat. Commun.* **2020**, *11* (1), 5990.

[24]   Jeong, J. W.; Yeo, W. H.; Akhtar, A.; et al. Materials and optimized designs for human-machine interfaces via epidermal electronics. *Adv. Mater.* **2013**, *25* (47), 6839–6846.

[25]   Fan, J. A.; Yeo, W. H.; Su, Y.; et al. Fractal design concepts for stretchable electronics. *Nat. Commun.* **2014**, *5* (1), 3266.

[26]   Won, P.; Park, J. J.; Lee, T.; et al. Stretchable and transparent kirigami conductor of nanowire percolation network for electronic skin applications. *Nano Lett.* **2019**, *19* (9), 6087–6096.

[27]   Zhang, Z.; Yu, Y.; Tang, Y.; et al.Kirigami-inspired stretchable conjugated electronics. *Adv. Electron. Mater.* **2020**, *6* (1), 1900929.

[28]   Hwang, D. G.; Bartlett, M. D. Tunable mechanical metamaterials through hybrid kirigami structures. *Sci. Rep.* **2018**, *8* (1), 3378.

[29]   Li, B. M.; Kim, I.; Zhou, Y.; Mills, A. C.; Flewwellin, T. J.; Jur, J. S. Kirigami-inspired textile electronics: KITE. *Adv. Mater. Technol.* **2019**, *4* (11), 1900511.

[30] Rafsanjani, A.; Zhang, Y.; Liu, B.; Rubinstein, S. M.; Bertoldi, K. Kirigami skins make a simple soft actuator crawl. *Sci. Robot.* **2018**, *3* (15), eaar7555.

[31] Tang, Y.; Li, Y.; Hong, Y.; Yang, S.; Yin, J. Programmable active kirigami metasheets with more freedom of actuation. *Proc. Natl. Acad. Sci.* **2019**, *116* (52), 26407–26413.

[32] Li, Y.; Zhang, Q.; Hong, Y.; Yin, J. 3D transformable modular Kirigami based programmable metamaterials. *Adv. Funct. Mater.* **2021**, *31* (43), 2105641.

[33] Someya, T.; Bao, Z.; Malliaras, G. G. The rise of plastic bioelectronics. *Nature* **2016**, *540* (7633), 379–385.

[34] Rafeedi, T.; Lipomi, D. J. Multiple pathways to stretchable electronics. *Science* **2022**, *378* (6625), 1174–1175.

[35] Jiang, Y.; Zhang, Z.; Wang, Y. X.; et al. Topological supramolecular network enabled high-conductivity, stretchable organic bioelectronics. *Science* **2022**, *375* (6587), 1411–1417.

[36] Xu, J.; Wang, S.; Wang, GJN.; et al. Highly stretchable polymer semiconductor films through the nanoconfinement effect. *Science* **2017**, *355* (6320), 59–64.

[37] Yang, C.; Suo, Z. Hydrogel ionotronics. *Nat. Rev. Mater.* **2018**, *3* (6), 125–142.

[38] Wang, M.; Zhang, P.; Shamsi, M.; et al. Tough and stretchable ionogels by in situ phase separation. *Nat. Mater.* **2022**, *21* (3), 359–365.

[39] Bai, Y.; Chen, B.; Xiang, F.; Zhou, J.; Wang, H.; Suo, Z. Transparent hydrogel with enhanced water retention capacity by introducing highly hydratable salt. *Appl. Phys. Lett.* **2014**, *105* (15), 151903.

[40] Carpenter, J. A.; Eberle, T. B.; Schuerle, S.; Rafsanjani, A.; Studart, A. R. Facile manufacturing route for magneto-responsive soft actuators. *Adv. Intel. Syst.* **2021**, *3* (8), 2000283.

[41] Gao, J. Y.; Chen, S.; Liu, T. Y.; Ye, J.; Liu, J. Additive manufacture of low melting point metal porous materials: Capabilities, potential applications and challenges. *Mater. Today* **2021**, *49* 201–230.

[42] Lazarus, N.; Meyer, C. D.; Bedair, S. S.; Slipher, G. A.; Kierzewski, I. M. Magnetic elastomers for stretchable inductors. *ACS Appl. Mater. Interfaces* **2015**, *7* (19), 10080–10084.

[43] Sharma, N.; Nair, N. M.; Nagasarvari, G.; Ray, D.; Swaminathan, P. A review of silver nanowire-based composites for flexible electronic applications. *Flex. Print. Electron.* **2022**, *7*, 014009.

[44] Park, S.; Vosguerichian, M.; Bao, Z. A review of fabrication and applications of carbon nanotube film-based flexible electronics. *Nanoscale* **2013**, *5* (5), 1727–1752.

[45] Bhanushali, S.; Ghosh, P. C.; Simon, G. P.; Cheng, W. Copper nanowire-filled soft elastomer composites for applications as thermal interface materials. *Adv. Mater. Interfaces* **2017**, *4* (17), 1700387.

[46] Chen, S.; Deng, Z.; Liu, J. High performance liquid metal thermal interface materials. *Nanotechnology* **2020**, *32* (9), 092001.

[47] Novikov, A.; Goding, J.; Chapman, C.; Cuttaz, E.; Green, R. A. Stretchable bioelectronics: Mitigating the challenges of the percolation threshold in conductive elastomers. *APL Mater.* **2020**, *8* (10), 101105.

[48] Sun, H.; Han, Z.; Willenbacher, N. Ultrastretchable conductive elastomers with a low percolation threshold for printed soft electronics. *ACS Appl. Mater. Interfaces* **2019**, *11* (41), 38092–38102.

[49] Kaur, G.; Adhikari, R.; Cass, P.; Bown, M.; Gunatillake, P. Electrically conductive polymers and composites for biomedical applications. *RSC Adv.* **2015**, *5* (47), 37553–37567.

[50] Tutika, R.; Zhou, S. H.; Napolitano, R. E.; Bartlett, M. D. Mechanical and functional tradeoffs in multiphase liquid metal, solid particle soft composites. *Adv. Funct. Mater.* **2018**, *28* (45), 1804336.

[51] Style, R. W.; Tutika, R.; Kim, J. Y.; Bartlett, M. D. Solid–liquid composites for soft multifunctional materials. *Adv. Funct. Mater.* **2021**, *31* (1), 2005804.

[52] Haque, A. T.; Tutika, R.; Byrum, R. L.; Bartlett, M. D. Programmable liquid metal microstructures for multifunctional soft thermal composites. *Adv. Funct. Mater.* **2020**, *30* (25), 2000832.

[53] Ford, M. J.; Patel, D. K.; Pan, C.; Bergbreiter, S.; Majidi, C. Controlled assembly of liquid metal inclusions as a general approach for multifunctional composites. *Adv. Mater.* **2020**, *32* (46), 2002929.

[54] Schubert, B. E.; Floreano, D. Variable stiffness material based on rigid low-melting-point-alloy microstructures embedded in soft poly (dimethylsiloxane)(PDMS). *RSC Adv.* **2013**, *3* (46), 24671–24679.

[55] Sharifi, S.; Mohammadi Nasab, A.; Chen, P. E.; Liao, Y.; Jiao, Y.; Shan, W. Robust Bicontinuous Elastomer–Metal Foam Composites with Highly Tunable Stiffness. *Adv. Eng. Mater.* **2022**, *24* (8), 2101533.

[56] Barron III, E. J.; Peterson, R. S.; Lazarus, N.; Bartlett, M. D. Mechanically cloaked multiphase magnetic elastomer soft composites for wearable wireless power transfer. *ACS Appl. Mater. Interfaces* **2020**, *12* (45), 50909–50917.

[57] Testa, P.; Style, R. W.; Cui, J.; et al. Magnetically Addressable Shape-Memory and Stiffening in a Composite Elastomer. *Adv. Mater.* **2019**, *31* (29), 1900561.

[58] Bastola, A.; Hoang, V.; Li, L. A novel hybrid magnetorheological elastomer developed by 3D printing. *Mater. Des.* **2017**, *114*, 391–397.

[59] Bartlett, M. D.; Kazem, N.; Powell-Palm, M. J.; et al. High thermal conductivity in soft elastomers with elongated liquid metal inclusions. *Proc. Natl. Acad. Sci.* **2017**, *114* (9), 2143–2148.

[60] Hwang, D.; Barron III, E. J.; Haque, A. T.; Bartlett, M. D. Shape morphing mechanical metamaterials through reversible plasticity. *Sci. Robot.* **2022**, *7* (63), eabg2171.

[61] Terryn, S.; Brancart, J.; Lefeber, D.; Van Assche, G.; Vanderborght, B. Self-healing soft pneumatic robots. *Sci. Robot.* **2017**, *2* (9), eaan4268.

[62] Tonazzini, A.; Mintchev, S.; Schubert, B.; Mazzolai, B.; Shintake, J.; Floreano, D. Variable stiffness fiber with self-healing capability. *Adv. Mater.* **2016**, *28* (46), 10142–10148.

[63] Barron III, E. J.; Williams, E. T.; Tutika, R.; Lazarus, N.; Bartlett, M. D. A unified understanding of magnetorheological elastomers for rapid and extreme stiffness tuning. *RSC Appl. Polym.* **2023**, *1*, 315–324.

[64] Chu, A. H.; Cheng, T.; Muralt, A.; Onal, C. D. A passively conforming soft robotic gripper with three-dimensional negative bending stiffness fingers. *Soft Robot.* **2023**, *10* (3), 556–567.

[65] Tondu, B.; Lopez, P. Modeling and control of McKibben artificial muscle robot actuators. *IEEE Control Syst. Mag.* **2000**, *20* (2), 15–38.

[66] Suzumori, K.; Iikura, S.; Tanaka, H. Flexible microactuator for miniature robots. In: *[1991] proceedings. IEEE micro electro mechanical systems*; IEEE, 1991; pp. 204–209.

[67] Suzumori, K.; Iikura, S.; Tanaka, H. Applying a flexible microactuator to robotic mechanisms. *IEEE Control Syst. Mag.* **1992**, *12* (1), 21–27.

[68] Noritsugu, T.; Kubota, M.; Yoshimatsu, S. Development of pneumatic rotary soft actuator made of silicone rubber. *J. Robot. Mechatron.* **2001**, *13* (1), 17–22.

[69] Suzumori, K.; Asaad, S. A novel pneumatic rubber actuator for mobile robot bases. In *Proceedings of IEEE/RSJ international conference on intelligent robots and systems. IROS'96.* vol. 2; IEEE, 1996; pp. 1001–1006.

[70] Walker, J.; Zidek, T.; Harbel, C.; et al. Soft robotics: a review of recent developments of pneumatic soft actuators. In *Actuators.* vol. 9; MDPI, 2020, p. 3.

[71] Huang, W.; Xiao, J.; Xu, Z. A variable structure pneumatic soft robot. *Sci. Rep.* **2020**, *10* (1), 18778.

[72] Chen, S.; Tan, M. W. M., Gong X.; Lee, P. S. Low-voltage soft actuators for interactive human–machine interfaces. *Adv. Intel. Syst.* **2022**, *4* (2), 2100075.

[73] Chung, H. J.; Parsons, A. M.; Zheng, L. Magnetically controlled soft robotics utilizing elastomers and gels in actuation: a review. *Adv. Intel. Syst.* **2021**, *3* (3), 2000186.

[74] Wang, L.; Yang, Y.; Chen, Y.; et al. Controllable and reversible tuning of material rigidity for robot applications. *Mater. Today* **2018**, *21* (5), 563–576.

[75] Daeneke, T.; Khoshmanesh, K.; Mahmood, N.; et al. Liquid metals: fundamentals and applications in chemistry. *Chem. Soc. Rev.* **2018**, *47* (11), 4073–4111.

[76] Andrade, E. N. D. C.; Dobbs, E. The viscosities of liquid lithium, rubidium and caesium. *Proc. R. Soc. Lond. Ser. A, Math. Phys. Sci.* **1952**, *211* (1104), 12–30.

[77] Samsonov, G. The role of stable electron configurations in the origin of the properties of chemical elements and compounds. *Sov. Powder Metall. Metal Ceram.* **1966**, *5* (12), 955–963.

[78]   Dickey, M. D. Emerging applications of liquid metals featuring surface oxides. *ACS Appl. Mater. Interfaces* **2014**, *6* (21), 18369–18379.

[79]   Rustagi, N.; Singh, R. Mercury and health care. *Indian J. Occup. Environ. Med.* **2010**, *14* (2), 45.

[80]   Shanks, N.; Lambourne, A.; Morton, C.; Sanford, J. Comparison of accuracy of digital and standard mercury thermometers. *Br. Med. J., Clin. Res. Ed.* **1983**, *287* (6401), 1263.

[81]   Liu, H.; Yu, Y.; Wang, W.; et al. Novel contrast media based on the liquid metal gallium for in vivo digestive tract radiography: a feasibility study. *Biometals* **2019**, *32* 795–801.

[82]   Cheng, S.; Wu, Z. Microfluidic electronics. *Lab Chip* **2012**, *12* (16), 2782–2791.

[83]   Yu, S.; Kaviany M. Electrical, thermal, and species transport properties of liquid eutectic Ga-In and Ga-In-Sn from first principles. *J. Chem. Phys.* **2014**, *140* (6), 064303.

[84]   Zhu, J. Y.; Tang, S. Y.; Khoshmanesh, K.; Ghorbani, K. An integrated liquid cooling system based on galinstan liquid metal droplets. *ACS Appl. Mater. Interfaces* **2016**, *8* (3), 2173–2180.

[85]   Farrell, Z. J.; Tabor, C. Control of gallium oxide growth on liquid metal eutectic gallium/indium nanoparticles via thiolation. *Langmuir* **2018**, *34* (1), 234–240.

[86]   Neumann, T. V.; Kara, B.; Sargolzaeiaval, Y.; et al. Aerosol spray deposition of liquid metal and elastomer coatings for rapid processing of stretchable electronics. *Micromachines* **2021**, *12* (2), 146.

[87]   Lin, Y.; Gordon, O.; Khan, M. R.; Vasquez, N.; Genzer, J.; Dickey, M. D. Vacuum filling of complex microchannels with liquid metal. *Lab Chip* **2017**, *17* (18), 3043–3050.

[88]   Kim, K.; Ahn, J.; Jeong, Y.; Choi, J.; Gul, O.; Park, I. All-soft multiaxial force sensor based on liquid metal for electronic skin. *Micro Nano Syst. Lett.* **2021**, *9* (1), 1–8.

[89]   Ladd, C.; So, J. H.; Muth, J.; Dickey, M. D. 3D printing of free standing liquid metal microstructures. *Adv. Mater.* **2013**, *25* (36), 5081–5085.

[90]   Bartlett, M. D.; Fassler, A.; Kazem, N.; Markvicka, E. J.; Mandal, P.; Majidi, C. Stretchable, high-k dielectric elastomers through liquid-metal inclusions. *Adv. Mater.* **2016**, *28* (19), 3726–3731.

[91]   Kazem, N.; Hellebrekers, T.; Majidi, C. Soft multifunctional composites and emulsions with liquid metals. *Adv. Mater.* **2017**, *29* (27), 1605985.

[92]   Krings, E. J.; Zhang, H.; Sarin, S.; Shield, J. E.; Ryu, S.; Markvicka, E. J. Lightweight, thermally conductive liquid metal elastomer composite with independently controllable thermal conductivity and density. *Small* **2021**, *17* (52), 2104762.

[93]   Tutika, R.; Kmiec, S.; Haque, A. T.; Martin, S. W.; Bartlett, M. D. Liquid metal–elastomer soft composites with independently controllable and highly tunable droplet size and volume loading. *ACS Appl. Mater. Interfaces* **2019**, *11* (19), 17873–17883.

[94]   Ford, M. J.; Palaniswamy, M.; Ambulo, C. P.; Ware, T. H.; Majidi, C. Size of liquid metal particles influences actuation properties of a liquid crystal elastomer composite. *Soft Matter* **2020**, *16* (25), 5878–5885.

[95]   Malakooti, M. H.; Bockstaller, M. R.; Matyjaszewski, K.; Majidi, C. Liquid metal nanocomposites. *Nanoscale Adv.* **2020**, *2* (7), 2668–2677.

[96]   Moon, S.; Kim, H.; Lee, K.; Park, J.; Kim, Y.; Choi, S. Q. 3D printable concentrated liquid metal composite with high thermal conductivity. *iScience* **2021**, *24* (10), 103183.

[97]   Pozarycki, T. A.; Hwang, D.; Barron III, E. J.; Wilcox, B. T.; Tutika, R.; Bartlett, M. D. Tough bonding of liquid metal-elastomer composites for multifunctional adhesives. *Small* **2022**, *18* (41), 2203700.

[98]   Haake, A.; Tutika, R.; Schloer, G. M.; Bartlett, M. D.; Markvicka, E. J. On-demand programming of liquid metal-composite microstructures through direct ink write 3D printing. *Adv. Mater.* **2022**, *34* (20), 2200182.

[99]   Hur, O.; Tutika, R.; Klemba, N.; Markvicka, E. J.; Bartlett, M. D. Designing liquid metal microstructures through directed material extrusion additive manufacturing. *Addit. Manuf.* **2024**, *79*, 103925.

[100]   Tutika, R.; Haque, A. T.; Bartlett, M. D. Self-healing liquid metal composite for reconfigurable and recyclable soft electronics. *Commun. Mater.* **2021**, *2* (1), 64.

[101]   Ning, N.; Huang, W.; Liu, S.; et al. Highly stretchable liquid metal/polyurethane sponge conductors with excellent electrical conductivity stability and good mechanical properties. *Composites, Part B, Eng.* **2019** (179), 107492.

[102] Boley, J. W.; White, E. L.; Kramer, R. K. Mechanically sintered gallium–indium nanoparticles. *Adv. Mater.* **2015**, *27* (14), 2355–2360.

[103] Thrasher, C. J.; Farrell, Z. J.; Morris, N. J.; Willey, C. L.; Tabor, C. E. Mechanoresponsive polymerized liquid metal networks. *Adv. Mater.* **2019**, *31* (40), 1903864.

[104] Boley, J. W.; Van Rees, W. M.; Lissandrello, C.; et al. Shape-shifting structured lattices via multimaterial 4D printing. *Proc. Natl. Acad. Sci.* **2019**, *116* (42), 20856–20862.

[105] Markvicka, E. J.; Tutika, R.; Bartlett, M. D.; Majidi, C. Soft electronic skin for multi-site damage detection and localization. *Adv. Funct. Mater.* **2019**, *29* (29), 1900160.

[106] Kim, S.; Kim, S.; Hong, K.; Dickey, M. D.; Park, S. Liquid-metal-coated magnetic particles toward writable, nonwettable, stretchable circuit boards, and directly assembled liquid metal-elastomer conductors. *ACS Appl. Mater. Interfaces* **2022**, *14* (32), 37110–37119.

[107] Yao, B.; Hong, W.; Chen, T.; et al. Highly stretchable polymer composite with strain-enhanced electromagnetic interference shielding effectiveness. *Adv. Mater.* **2020**, *32* (14), 1907499.

[108] Liang, S.; Li, Y.; Chen, Y.; et al. Liquid metal sponges for mechanically durable, all-soft, electrical conductors. *J. Mater. Chem. C* **2017**, *5* (7), 1586–1590.

[109] Zu, W.; Ohm, Y.; Carneiro, M. R.; Vinciguerra, M.; Tavakoli, M.; Majidi, C. A comparative study of silver microflakes in digitally printable liquid metal embedded elastomer inks for stretchable electronics. *Adv. Mater. Technol.* **2022**, *7* (12), 2200534.

[110] Zu, W.; Carranza, H. E.; Bartlett, M. D. Enhancing electrical conductivity of stretchable liquid metal–silver composites through direct ink writing. *ACS Appl. Mater. Interfaces* **2024**, *16* (18), 23895–23903.

[111] Lee, G. H.; Woo, H.; Yoon, C.; et al. A Personalized Electronic Tattoo for Healthcare Realized by On-the-Spot Assembly of an Intrinsically Conductive and Durable Liquid-Metal Composite. *Adv. Mater.* **2022**, *34* (32), 2204159.

[112] Zhu, L.; Wang, B.; Handschuh-Wang, S.; Zhou, X. Liquid metal–based soft microfluidics. *Small* **2020**, *16* (9), 1903841.

[113] Bark, H.; Tan, M. W. M.; Thangavel, G.; Lee, P. S. Deformable high loading liquid metal nanoparticles composites for thermal energy management. *Adv. Energy Mater.* **2021**, *11* (35), 2101387.

[114] Ralphs, M. I.; Kemme, N.; Vartak, P. B.; et al. In situ alloying of thermally conductive polymer composites by combining liquid and solid metal microadditives. *ACS Appl. Mater. Interfaces* **2018**, *10* (2), 2083–2092.

[115] Uppal, A.; Kong, W.; Rana, A.; et al. Precuring matrix viscosity controls thermal conductivity of elastomeric composites with compression-activated liquid and solid metallic filler networks. *Adv. Mater. Interfaces* **2023**, *10* (9), 2201875.

[116] Won, P.; Jeong, S.; Majidi, C.; Ko, S. H. Recent advances in liquid-metal-based wearable electronics and materials. *iScience* **2021** (24), 7.

[117] Hoang, T. T.; Phan, P. T.; Thai, M. T.; et al. Magnetically engineered conductivity of soft liquid metal composites for robotic, wearable electronic, and medical applications. *Adv. Intel. Syst.* **2022**, *4* (12), 2200282.

[118] Won, P.; Valentine, C. S.; Zadan, M.; et al. 3D printing of liquid metal embedded elastomers for soft thermal and electrical materials. *ACS Appl. Mater. Interfaces* **2022**, *14* (49), 55028–55038.

[119] Pan, C.; Liu, D.; Ford, M. J.; Majidi, C. Ultrastretchable, wearable triboelectric nanogenerator based on sedimented liquid metal elastomer composite. *Adv. Mater. Technol.* **2020**, *5* (11), 2000754.

[120] Zhu, L.; Chen, Y.; Shang, W.; et al. Anisotropic liquid metal–elastomer composites. *J. Mater. Chem. C* **2019**, *7* (33), 10166–10172.

[121] Palleau, E.; Reece, S.; Desai, S. C.; Smith, M. E.; Dickey, M. D. Self-healing stretchable wires for reconfigurable circuit wiring and 3D microfluidics. *Adv. Mater.* **2013**, *25* (11), 1589–1592.

[122] Krisnadi, F.; Nguyen, L. L.; Ankit; et al. Directed assembly of liquid metal–elastomer conductors for stretchable and self-healing electronics. *Adv. Mater.* **2020**, *32* (30), 2001642.

[123] Hao, Y.; Gao, J.; Lv, Y.; Liu, J. Low melting point alloys enabled stiffness tunable advanced materials. *Adv. Funct. Mater.* **2022**, *32* (25), 2201942.

[124] Buckner, T. L.; Yuen, M. C.; Kim, S. Y.; Kramer-Bottiglio, R. Enhanced variable stiffness and variable stretchability enabled by phase-changing particulate additives. *Adv. Funct. Mater.* **2019**, *29* (50), 1903368.

[125] Lipchitz, A.; Harvel, G.; Sunagawa, T. Thermophysical characteristics of liquid metal In-Bi-Sn eutectic (Field's metal) as a similarity coolant. *J. Nucl. Eng. Radiat. Sci.* **2022**, *8* (3), 031301.

[126] Yang, T.; Kang, J. G.; Weisensee, P. B.; et al. A composite phase change material thermal buffer based on porous metal foam and low-melting-temperature metal alloy. *Appl. Phys. Lett.* **2020**, *116* (7), 071901.

[127] Brunetti, B.; Gozzi, D.; Iervolino, M.; et al. Bismuth activity in lead-free solder Bi–In–Sn alloys. *Calphad* **2006**, *30* (4), 431–442.

[128] Yu, B.; Liu, L.; Liu, B.; Zhao, X.; Deng, W. Printing of low-melting-point alloy as top electrode for organic solar cells. *Adv. Opt. Mater.* **2023**, *11* (2), 2201977.

[129] Shan, W.; Lu, T.; Majidi, C. Soft-matter composites with electrically tunable elastic rigidity. *Smart Mater. Struct.* **2013**, *22* (8), 085005.

[130] Mohammadi Nasab, A.; Buckner, T. L.; Yang, B.; Kramer-Bottiglio, R. Effect of filler aspect ratio on stiffness and conductivity in phase-changing particulate composites. *Adv. Mater. Technol.* **2022**, *7* (5), 2100920.

[131] Buckner, T. L.; Yuen, M. C.; Kramer-Bottiglio, R. Shape memory silicone using phase-changing inclusions. In *2020 3rd IEEE international conference on soft robotics (RoboSoft)*; IEEE, 2020; pp. 259–265.

[132] Pashine, N.; Nasab, A. M.; Kramer-Bottiglio, R. Reprogrammable allosteric metamaterials from disordered networks. *Soft Matter* **2023**, *19* (8), 1617–1623.

[133] Allen, E. A.; Swensen, J. P. Directional stiffness control through geometric patterning and localized heating of Field's metal lattice embedded in silicone. In *Actuators*. vol. 7; MDPI, 2018, p. 80.

[134] Bhuyan, P.; Wei, Y.; Sin, D.; et al. Soft and stretchable liquid metal composites with shape memory and healable conductivity. *ACS Appl. Mater. Interfaces* **2021**, *13* (24), 28916–28924.

[135] Takahashi, R.; Sun, T. L.; Saruwatari, Y.; Kurokawa, T.; King, D. R.; Gong, J. P. Creating stiff, tough, and functional hydrogel composites with low-melting-point alloys. *Adv. Mater.* **2018**, *30* (16), 1706885.

[136] Zhalmuratova, D.; Chung, H. J. Reinforced gels and elastomers for biomedical and soft robotics applications. *ACS Appl. Polym. Mater.* **2020**, *2* (3), 1073–1091.

[137] Deng, F.; Nguyen, Q. K.; Zhang, P. Multifunctional liquid metal lattice materials through hybrid design and manufacturing. *Addit. Manuf.* **2020**, *33*, 101117.

[138] Long, F.; Shao, Y.; Zhao, Z.; et al. Printable multi-stage variable stiffness material enabled by low melting point particle additives. *J. Mater. Chem. C* **2023**, *11*, 1285–1297.

[139] Van Meerbeek, I. M.; Mac Murray, B. C.; Kim, J. W.; et al. Morphing metal and elastomer bicontinuous foams for reversible stiffness, shape memory, and self-healing soft machines. *Adv. Mater.* **2016**, *28* (14), 2801–2806.

[140] Tevis, I. D.; Newcomb, L. B.; Thuo, M. Synthesis of liquid core–shell particles and solid patchy multicomponent particles by shearing liquids into complex particles (SLICE). *Langmuir* **2014**, *30* (47), 14308–14313.

[141] Idrus-Saidi, S. A.; Tang, J.; Ghasemian, M. B.; et al. Liquid metal core–shell structures functionalised via mechanical agitation: the example of Field's metal. *J. Mater. Chem. A* **2019**, *7* (30), 17876–17887.

[142] Han, Z.; Yang, B.; Qi, Y.; Cumings, J. Synthesis of low-melting-point metallic nanoparticles with an ultrasonic nanoemulsion method. *Ultrasonics* **2011**, *51* (4), 485–488.

[143] Chang, B. S.; Tutika, R.; Cutinho, J.; et al. Mechanically triggered composite stiffness tuning through thermodynamic relaxation (ST3R). *Mater. Horiz.* **2018**, *5* (3), 416–422.

[144] Park, S.; Baugh, N.; Shah, H. K.; Parekh, D. P.; Joshipura, I. D.; Dickey, M. D. Ultrastretchable elastic shape memory fibers with electrical conductivity. *Adv. Sci.* **2019**, *6* (21), 1901579.

[145] Peters, J.; Nolan, E.; Wiese, M.; et al. Actuation and stiffening in fluid-driven soft robots using low-melting-point material. In *2019 IEEE/RSJ international conference on intelligent robots and systems (IROS)*; IEEE, 2019, pp. 4692–4698.

[146] Xu, Y.; Chen, P. E.; Li, H.; et al. Correlation-function-based microstructure design of alloy-polymer composites for dynamic dry adhesion tuning in soft gripping. *J. Appl. Phys.* **2022**, *131*, 11.

[147] Mohammadi Nasab, A.; Stampfli, P.; Sharifi, S.; Luo, A.; Turner, K. T.; Shan, W. Dynamically tunable dry adhesion through a subsurface thin layer with tunable stiffness. *Adv. Mater. Interfaces* **2022**, *9* (7), 2102080.

[148] Han, Y.; Dong, J. Electrohydrodynamic (EHD) printing of molten metal ink for flexible and stretchable conductor with self-healing capability. *Adv. Mater. Technol.* **2018**, *3* (3), 1700268.

[149] De Vicente, J.; Klingenberg, D. J.; Hidalgo-Alvarez, R. Magnetorheological fluids: a review. *Soft Matter* **2011**, *7* (8), 3701–3710.

[150] Bossis, G.; Lacis, S.; Meunier, A.; Volkova, O. Magnetorheological fluids. *J. Magn. Magn. Mater.* **2002**, *252*, 224–228.

[151] Ashtiani, M.; Hashemabadi, S.; Ghaffari, A. A review on the magnetorheological fluid preparation and stabilization. *J. Magn. Magn. Mater.* **2015**, *374*, 716–730.

[152] Zhu, X.; Jing, X.; Cheng, L. Magnetorheological fluid dampers: a review on structure design and analysis. *J. Intell. Mater. Syst. Struct.* **2012**, *23* (8), 839–873.

[153] Lai, C. Y.; Liao, W. H. Vibration control of a suspension system via a magnetorheological fluid damper. *J. Vib. Control* **2002**, *8* (4), 527–547.

[154] Klingenberg, D. J. Magnetorheology: applications and challenges. *AIChE J.* **2001**, *47* (2), 246.

[155] Felt, D. W.; Hagenbuchle, M.; Liu, J.; Richard, J. Rheology of a magnetorheological fluid. *J. Intell. Mater. Syst. Struct.* **1996**, *7* (5), 589–593.

[156] Lazarus, N.; Bedair, S. Improved power transfer to wearable systems through stretchable magnetic composites. *Appl. Phys. A* **2016**, *122*, 1–7.

[157] Silva, J.; Gouveia, C.; Dinis, G.; Pinto, A.; Pereira, A. Giant magnetostriction in low-concentration magnetorheological elastomers. *Composites, Part B, Eng.* **2022** (243), 110125.

[158] Rohim, M. A. S.; Nazmi, N.; Bahiuddin, I.; Mazlan, S. A.; Nordin, N. A. A mini review on modeling magnetostriction behavior of magnetorheological solid materials. In *2022 IEEE 12th symposium on computer applications & industrial electronics (ISCAIE)*; IEEE, 2022, pp. 133–138.

[159] Li, W.; Zhang, X. A study of the magnetorheological effect of bimodal particle based magnetorheological elastomers. *Smart Mater. Struct.* **2010**, *19* (3), 035002.

[160] Boczkowska, A.; Awietjan, S. F.; Wroblewski, R. Microstructure–property relationships of urethane magnetorheological elastomers. *Smart Mater. Struct.* **2007**, *16* (5), 1924.

[161] Kashima, S.; Miyasaka, F.; Hirata, K. Novel soft actuator using magnetorheological elastomer. *IEEE Trans. Magn.* **2012**, *48* (4), 1649–1652.

[162] Böse, H.; Rabindranath, R.; Ehrlich, J. Soft magnetorheological elastomers as new actuators for valves. *J. Intell. Mater. Syst. Struct.* **2012**, *23* (9), 989–994.

[163] Sun, S.; Yang, J.; Li, W.; et al. Development of an isolator working with magnetorheological elastomers and fluids. *Mech. Syst. Signal Process.* **2017**, *83*, 371–384.

[164] Liao, G.; Gong, X.; Xuan, S.; Kang, C.; Zong, L. Development of a real-time tunable stiffness and damping vibration isolator based on magnetorheological elastomer. *J. Intell. Mater. Syst. Struct.* **2012**, *23* (1), 25–33.

[165] Carlson, J. D.; Jolly, M. R. MR fluid, foam and elastomer devices. *Mechatronics* **2000**, *10* (4–5), 555–569.

[166] Böse, H.; Röder, R. Magnetorheological elastomers with high variability of their mechanical properties. *J. Phys. Conf. Ser.* **2009**, *149*, 012090.

[167] Tong, Y.; Dong, X.; Qi, M. Improved tunable range of the field-induced storage modulus by using flower-like particles as the active phase of magnetorheological elastomers. *Soft Matter* **2018**, *14* (18), 3504–3509.

[168] Ginder, J. M.; Nichols, M. E.; Elie, L. D.; Tardiff, J. L. Magnetorheological elastomers: properties and applications. In *Smart structures and materials 1999: smart materials technologies*. vol. 3675; SPIE, 1999, pp. 131–138.
[169] Manti, M.; Cacucciolo, V.; Cianchetti, M. Stiffening in soft robotics: A review of the state of the art. *IEEE Robot. Autom. Mag.* **2016**, *23* (3), 93–106.
[170] Ginder, J. M.; Nichols, M. E.; Elie, L. D.; Clark, S. M. Controllable-stiffness components based on magnetorheological elastomers. In *Smart structures and materials 2000: smart structures and integrated systems*. vol. 3985; SPIE, 2000, pp. 418–425.
[171] Li, Y.; Li, J.; Li, W.; Du, H. A state-of-the-art review on magnetorheological elastomer devices. *Smart Mater. Struct.* **2014**, *23* (12), 123001.
[172] Kang, S. S.; Choi, K.; Nam, J. D.; Choi, H. J. Magnetorheological elastomers: Fabrication, characteristics, and applications. *Materials* **2020**, *13* (20), 4597.
[173] Shuib, R. K.; Pickering, K. L.; Mace, B. R. Dynamic properties of magnetorheological elastomers based on iron sand and natural rubber. *J. Appl. Polym. Sci.* **2015**, *132*, 8.
[174] Yu, M.; Yang, P.; Fu, J.; Liu, S.; Qi, S. Study on the characteristics of magneto-sensitive electromagnetic wave-absorbing properties of magnetorheological elastomers. *Smart Mater. Struct.* **2016**, *25* (8), 085046.
[175] Wang, Y.; Zhang, X.; Oh, J.; Chung, K. Fabrication and properties of magnetorheological elastomers based on CR/ENR self-crosslinking blends. *Smart Mater. Struct.* **2015**, *24* (9), 095006.
[176] Cestarollo, L.; Smolenski, S.; El-Ghazaly, A. Nanoparticle-based magnetorheological elastomers with enhanced mechanical deflection for haptic displays. *ACS Appl. Mater. Interfaces* **2022**, *14* (16), 19002–19011.
[177] Dobroserdova, A.; Schümann, M.; Borin, D.; Novak, E.; Odenbach, S.; Kantorovich, S. Magneto-elastic coupling as a key to microstructural response of magnetic elastomers with flake-like particles. *Soft Matter* **2022**, *18* (3), 496–506.
[178] Dyniewicz, B.; Bajkowski, J. M.; Bajer, C. I. Semi-active control of a sandwich beam partially filled with magnetorheological elastomer. *Mech. Syst. Signal Process.* **2015**, *60*, 695–705.
[179] Zhou, Y.; Jerrams, S.; Betts, A.; Farrell, G.; Chen, L. The influence of particle content on the equi-biaxial fatigue behaviour of magnetorheological elastomers. *Mater. Des.* **2015**, *67*, 398–404.
[180] Zhang, J.; Qiao, Y.; Zhang, M.; Zhai, P. Magnetorheological behavior of isotropic silicone rubber-based magnetorheological elastomers under coupled static–dynamic compressive loads. *Smart Mater. Struct.* **2022**, *31* (9), 095010.
[181] Kallio, M.; Lindroos, T.; Aalto, S.; Järvinen, E.; Kärnä, T.; Meinander, T. Dynamic compression testing of a tunable spring element consisting of a magnetorheological elastomer. *Smart Mater. Struct.* **2007**, *16* (2), 506.
[182] Bastola, A. K.; Paudel, M.; Li, L. Magnetic circuit analysis to obtain the magnetic permeability of magnetorheological elastomers. *J. Intell. Mater. Syst. Struct.* **2018**, *29* (14), 2946–2953.
[183] Chen, L.; Gong, X.; Li, W. Microstructures and viscoelastic properties of anisotropic magnetorheological elastomers. *Smart Mater. Struct.* **2007**, *16* (6), 2645.
[184] Boczkowska, A.; Awietjan, S.; Wejrzanowski, T.; Kurzydłowski, K. Image analysis of the microstructure of magnetorheological elastomers. *J. Mater. Sci.* **2009**, *44*, 3135–3140.
[185] Vatandoost, H.; Sedaghati, R.; Rakheja, S. Development of new phenomenological models for predicting magnetic permeability of isotropic and anisotropic magneto-rheological elastomers. *IEEE Trans. Instrum. Meas.* **2022**, *71*, 1–10.
[186] Bastola, A. K.; Li, L.; Paudel, M. A hybrid magnetorheological elastomer developed by encapsulation of magnetorheological fluid. *J. Mater. Sci.* **2018**, *53* 7004–7016.
[187] Testa, P.; Chappuis, B.; Kistler, S.; Style, R. W.; Heyderman, L. J.; Dufresne, E. R. Switchable adhesion of soft composites induced by a magnetic field. *Soft Matter* **2020**, *16* (25), 5806–5811.
[188] Sihvola, A. H. *Electromagnetic mixing formulas and applications. IEE electromagnetic waves series*. Institution of Electrical Engineers, 1999. Available from: https://books.google.com/books?id=uIHSNwxBxjgC.

[189] Waki, H.; Igarashi, H.; Honma, T. Estimation of effective permeability of magnetic composite materials. *IEEE Trans. Magn.* **2005**, *41* (5), 1520–1523.

[190] Style, R. W.; Boltyanskiy, R.; Allen, B.; et al. Stiffening solids with liquid inclusions. *Nat. Phys.* **2015**, *11* (1), 82–87.

[191] Bira, N.; Dhagat, P.; Davidson, J. R. A review of magnetic elastomers and their role in soft robotics. *Front. Robot. AI* **2020** (7), 588391.

[192] Hua, D.; Liu, X.; Sun, S.; Sotelo, M. A.; Li, Z.; Li, W. A magnetorheological fluid-filled soft crawling robot with magnetic actuation. *IEEE/ASME Trans. Mechatron.* **2020**, *25* (6), 2700–2710.

[193] Qi, S.; Guo, H.; Fu, J.; Xie, Y.; Zhu, M.; Yu, M. 3D printed shape-programmable magneto-active soft matter for biomimetic applications. *Compos. Sci. Technol.* **2020** (188), 107973.

[194] Kim, J.; Chung, S. E.; Choi, S. E.; Lee, H.; Kim, J.; Kwon, S. Programming magnetic anisotropy in polymeric microactuators. *Nat. Mater.* **2011**, *10* (10), 747–752.

[195] Sun, M.; Tian, C.; Mao, L.; et al. Reconfigurable magnetic slime robot: deformation, adaptability, and multifunction. *Adv. Funct. Mater.* **2022**, *32* (26), 2112508.

[196] Ramachandran, V.; Bartlett, M. D.; Wissman, J.; Majidi, C. Elastic instabilities of a ferroelastomer beam for soft reconfigurable electronics. *Extreme Mech. Lett.* **2016**, *9*, 282–290.

[197] Croll, A. B.; Hosseini, N.; Bartlett, M. D. Switchable adhesives for multifunctional interfaces. *Adv. Mater. Technol.* **2019**, *4* (8), 1900193.

[198] Lee, S.; Yim, C.; Kim, W.; Jeon, S. Magnetorheological elastomer films with tunable wetting and adhesion properties. *ACS Appl. Mater. Interfaces* **2015**, *7* (35), 19853–19856.

[199] Kovalev, A.; Belyaeva, I. A.; von Hofen, C.; Gorb, S.; Shamonin, M. Magnetically switchable adhesion and friction of soft magnetoactive elastomers. *Adv. Eng. Mater.* **2022**, *24* (10), 2200372.

[200] Choi, D. S.; Kim, T. H.; Lee, S. H.; Pang, C.; Bae, J. W.; Kim, S. Y. Beyond human hand: shape-adaptive and reversible magnetorheological elastomer-based robot gripper skin. *ACS Appl. Mater. Interfaces* **2020**, *12* (39), 44147–44155.

[201] Bira, N.; Dhagat, P.; Davidson, J. R. Tuning the grasping strength of soft actuators with magnetic elastomer fingertips. *Smart Mater. Struct.* **2022**, *31* (4), 045013.

[202] Hu, W.; Lum, G. Z.; Mastrangeli, M.; Sitti, M. Small-scale soft-bodied robot with multimodal locomotion. *Nature* **2018**, *554* (7690), 81–85.

[203] Kim, Y.; Yuk, H.; Zhao, R.; Chester, S. A.; Zhao, X. Printing ferromagnetic domains for untethered fast-transforming soft materials. *Nature* **2018**, *558* (7709), 274–279.

[204] Kim, Y.; Parada, G. A.; Liu, S.; Zhao, X. Ferromagnetic soft continuum robots. *Sci. Robot.* **2019**, *4* (33), eaax7329.

[205] Yun, G.; Tang, S. Y.; Zhao, Q.; et al. Liquid metal composites with anisotropic and unconventional piezoconductivity. *Matter* **2020**, *3* (3), 824–841.

[206] Zhang, M.; Chen, X.; Sun, Y.; et al. A magnetically and thermally controlled liquid metal variable stiffness material. *Adv. Eng. Mater.* **2023**, *25* (6), 2201296.

[207] Buckner, T. L.; Bilodeau, R. A.; Kim, S. Y.; Kramer-Bottiglio, R. Roboticizing fabric by integrating functional fibers. *Proc. Natl. Acad. Sci.* **2020**, *117* (41), 25360–25369.

[208] Booth, J. W.; Cyr-Choiniere, O.; Case, J. C.; Shah, D.; Yuen, M. C.; Kramer-Bottiglio, R. Surface actuation and sensing of a tensegrity structure using robotic skins. *Soft Robot.* **2021**, *8* (5), 531–541.

[209] Shah, D. S.; Booth, J. W.; Baines, R. L.; et al. Tensegrity robotics. *Soft Robot.* **2022**, *9* (4), 639–656.

[210] Wang, Q.; Pan, C.; Zhang, Y.; et al. Magnetoactive liquid-solid phase transitional matter. *Matter* **2023**, *6* (3), 855–872.

Young Min Song, Sehui Chang, and Ji-Eun Yeo

Stretchable photodetectors and image sensors

Abstract: Light detection provides a wealth of vital information collected from both the outside world and inside the body. Photodetectors serve as essential components in imaging systems and sensing devices by converting incident light signals to electrical signals. Beyond the limitations of conventional optoelectronic devices in rigid and planar platforms, novel fabrication strategies for stretchable photodetectors are necessary to address complex optical configurations in imaging systems and the current limited functional and structural freedom of wearable devices. To develop the next-generation imaging/sensing systems with high performance and compactness, research on flexible and stretchable photodetector arrays has been extensively conducted. Recent works focus on various flexible and stretchable photodetecting devices using functional materials and structural designs. In this chapter, major approaches to realize the flexible and stretchable photodetectors are discussed in terms of nanomaterials and their hybrids and structural engineering. The state-of-the-art imaging and sensing devices presented here highlight the remaining challenges and future directions in the development of flexible and stretchable photodetectors.

1 Introduction

Photodetectors (PDs) act as a primary component of imaging systems in modern electronics. With the growth of semiconductor technologies, the fabrication methods of PDs have matured significantly in terms of optical/electrical performance, cost-effectiveness, and commercialization. However, flexible and stretchable PDs have garnered great attention in both research and industry due to their high compatibility with other flexible electronics. Stretchable PDs can meet the demands of bending, twisting, and deforming into curvilinear or uneven forms, characteristics critically required for next-generation imaging systems (e. g., hemispherical artificial retinas and wearable devices). Several strategies have been explored to realize flexible and stretchable PDs using inorganic, organic, and hybrid semiconductor materials as active materials. For example, inorganic semiconductor materials exhibit high-level performance and stability, but their inherent stiffness and brittleness pose significant challenges for use in stretchable

Young Min Song, School of Electrical Engineering and Computer Science, AI Graduate School, Department of Semiconductor Engineering, Gwangju Institute of Science and Technology, Unit 207, EECS A, 123 Cheomdangwagi-ro, Buk-gu, Gwangju, 61005, Republic of Korea, e-mail: ymsong@gist.ac.kr
Sehui Chang, Ji-Eun Yeo, School of Electrical Engineering and Computer Science, Gwangju Institute of Science and Technology, Unit 211, EECS A, 123 Cheomdangwagi-ro, Buk-gu, Gwangju, 61005, Republic of Korea, e-mails: shchangj@gm.gist.ac.kr, jieunyeo@gm.gist.ac.kr

https://doi.org/10.1515/9783110757286-004

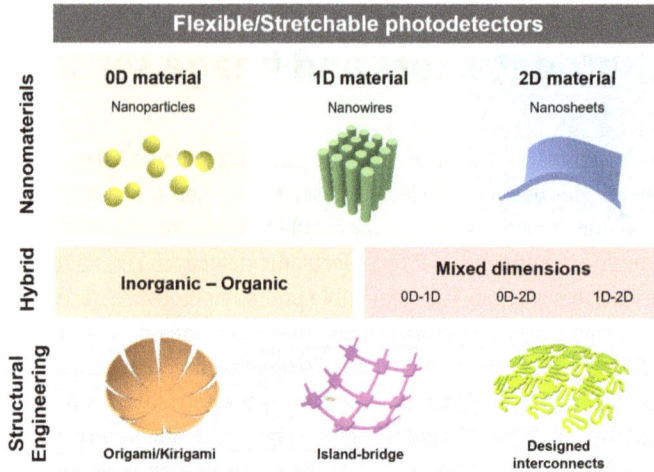

Figure 1: Overview of material and structural approaches for flexible/stretchable photodetectors. Reproduced with permission from Lee et al., Adv. Funct. Mater. 28, 1705202 (2018) [23].

devices. Nevertheless, research efforts continue in inorganic stretchable PDs, focusing not only on high-level electric performance but also on stretchability, utilizing various structural engineering techniques such as kirigami/origami [1–5] and island–bridge [6–9], and adopting nanomaterials (e. g., nanoparticles, nanowires (NWs), and nanomembranes (NMs)) and hybrid materials (e. g., inorganic–organic and mixed dimensional materials) (Figure 1) [10–19]. Simultaneously, organic PDs formed from small molecular or polymer materials have also been extensively explored due to their inherent flexibility, light-weight, solution-based fabrication methods, and facile tunability of optoelectronic properties [20–22]. Recent research on organic PDs focuses on improving performance and stability to compete with that of inorganic semiconductor materials. Additionally, hybrid approaches involving various materials, including inorganic–organic hybrids and mixed dimensional nanomaterials, offer another avenue for stretchable PDs by leveraging advantages from both types of materials.

An image sensor, or PD array, is an essential imaging component in optical camera systems that converts optical signals to electrical signals, allowing the visual information focused by the lens to be collected in electronic storage systems. Conventional camera systems suffer from physical limits in size reduction caused by multiple lens configurations needed to focus on the planar image plane. This issue arises because the lenses inherently form images with field curvature, which induces defocused images and low image sharpness [24]. In contrast, most animal eyes consist of a single spherical lens and a curvilinear retina—a much simpler optical structure with one or a few lenses compared to conventional camera systems [8, 9]. The visual accommodation to continuously focus on the objects at various distances can also be realized by modulating the curvature of the retina owing to its flexibility and stretchability. Hence, the development of flexible and stretchable PDs is highly in demand to realize miniaturized camera systems and advanced imaging devices. Moreover, recent progress in flexible displays, soft robotics, wearable devices, and biosensors also necessitates the integration of flexible and stretchable PDs to work with other electronic elements.

This chapter introduces recent studies with facile approaches for realizing flexible and stretchable PDs. Functional materials pave the way to fundamentally enhance the flexibility of devices, including nanodimensional materials (e. g., 0D nanoparticles, 1D nanowires, 2D nanomembranes, and mixed nanomaterials). Structural engineering of the device, e. g., using origami, kirigami, and island–bridge techniques, enables not only competitive optical and electrical performance compared to conventional planar devices but also high flexibility and stretchability. Moreover, state-of-the-art research in flexible and stretchable optoelectronic devices, such as image sensors, artificial vision systems, wearable devices, and bio-signal monitoring systems, identifies potential applications for flexible and stretchable PDs.

2 Flexible and stretchable photodetectors

With the rapid progress in flexible optoelectronic devices, PDs that offer both stretchability and high performance have been explored over the past decades. To realize flexible and stretchable PDs, the materials comprising the PDs (e. g., active materials, electrodes, and substrates) should be sufficiently flexible and stretchable while maintaining their photodetecting performance. In the case of substrates and electrodes, flexible substrates (e. g., polyethylene terephthalate (PET), polydimethylsiloxane (PDMS), mica, polyimide (PI), etc.) replace conventional rigid and bulk donor substrates, and several methods for fabricating flexible electrodes have also been devised, including thinning, using patterns such as serpentine and fractals, buckling, and controlling delamination [25]. Regarding active materials, the most common approach to achieve stretchability is the use of organic semiconductors that have an intrinsic ability to bend or stretch, compared to brittle and rigid inorganic semiconductors. On the other hand, although inorganic counterparts exhibit rigidity and brittleness, their performance in terms of responsivity, detectivity, and sensitivity is superior to that of organic semiconductors. For high-performance and flexible PDs, various strategies have been explored in inorganic, organic, and hybrid semiconductors using structural engineering techniques, such as kirigami, origami, and island–bridge, and by incorporating nanodimensional materials such as 0-dimensional (0D), 1-dimensional (1D), and 2-dimensional (2D). In this section, material and structural techniques for flexible and stretchable PDs will be introduced along with representative studies to date.

2.1 Nanomaterial-based flexible and stretchable PDs

2.1.1 0-dimensional (0D) materials-based flexible and stretchable photodetectors

Over the past decades, various efforts have been made to fabricate flexible PDs using 0-dimensional nanomaterials, such as quantum dots (QDs), nanoparticles (NPs), and nanocrystals (NCs). Zero-dimensional NPs are miniscule semiconducting particles with radii in the nanometer range that possess interesting optical and electrical properties. Additionally, their properties can be engineered by tuning the size and shape and by mixing various materials, such as altering optical absorption and emission bands within a broadband wavelength range from ultraviolet to terahertz frequencies, which is favorable for optoelectronic applications. Furthermore, the solution-phase synthesis of 0D nanoparticles is advantageous for device fabrication, enabling various solution-based processes including spin-coating and printing techniques that lead to decreased production costs. For 0D nanoparticle-based PDs, a variety of semiconducting materials is used in 0D nanoparticle forms, such as ZnO NPs [26], CdSe QDs [11], PbS QDs [27], Ag NCs [12], and CsPbBr$_3$ QDs [14, 28]. For example, Oertel et al. demonstrated CdSe QD PDs with a sandwich geometry in the visible wavelength range from 350 nm to 575 nm [11]. The solution-processed CdSe QD film has a thickness of 200 nm, which is a potential fabrication method for flexible PDs in extensive areas. Ultraviolet PDs were developed using an active layer of a blend of ZnO NPs and semiconducting polymers [10]. The PD structure consists of a polymer blend ZnO NPs active layer sandwiched with electrodes made of transparent indium tin oxide (ITO) and an aluminum. This new type of hybrid PD demonstrates a Schottky contact under dark conditions and an ohmic contact under illumination, due to the interfacial trap-controlled charge injection with a detectivity of 3.4×10^{15} Jones.

High-quality PbS QD offers various advantages, including solution processability [29], high molar absorption coefficients [30], tunable bandgaps across a wide range [31], and high stability. Nevertheless, a single-layer PbS QD PD exhibits limited performance such as large dark currents and slow response times. Furthermore, hybrid PbS PDs, proposed as an alternative, still suffer from insufficient overall performance enhancement. To overcome the challenges associated with single-layer and hybrid PbS QD PDs, a bilayer PbS-only QD flexible PD was demonstrated on a PI substrate [27]. The designed bilayer contains two layers: tetrabutylammonium iodide (TBAI) and 1, 2-ethanedithiol (EDT)-modified PbS QDs. By adopting this bilayer design, the fabricated PD exhibits a fast light response, high detectivity, broad dynamic range, and good responsivity, compared to conventional PdS QD PDs. The current-voltage (I-V) curve of the device with different materials (i. e., PbS-TBAI, PbS-EDT, and PbS-TBAI/PbS-EDT) under dark- and bright-light conditions was measured, indicating that the PbS-TBAI/PzbS-EDT-based PD has a higher photoresponse. This progress is attributed to the QD junction interface photocarrier separation under illumination and enhanced recombination under dark conditions. Addi-

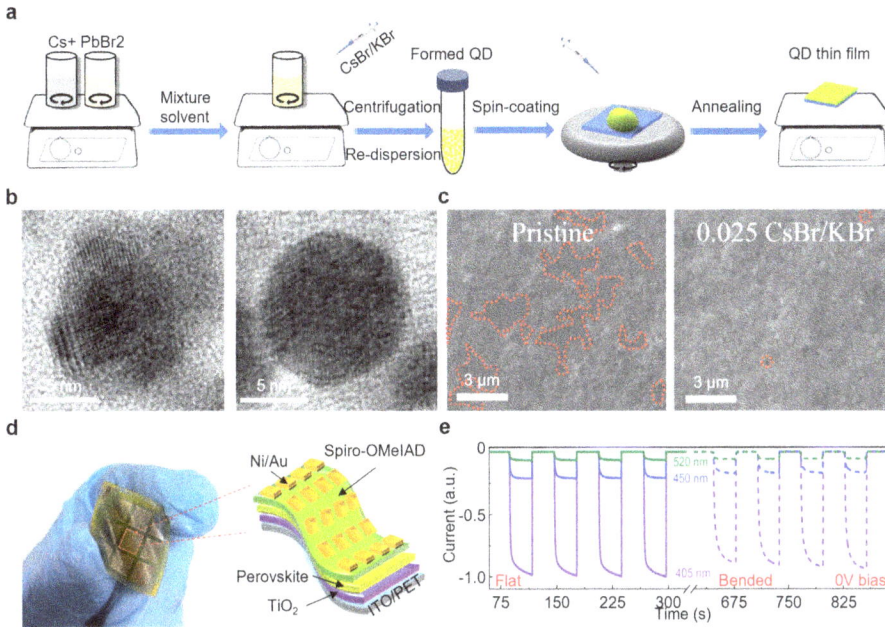

Figure 2: 0-dimensional material based flexible/stretchable PDs. (a) Illustration of all-inorganic perovskite (CsPbBr$_3$) QD synthesis process. (b) TEM image of synthesized colloidal CsPbBr$_3$ QDs without (left) and with the CsBr/KBr precursor (right). (c) The SEM images of the QD film surface morphology. (a)–(e) Reproduced with permission from Shen et al., Adv. Mater. 32, 2000004 (2020) [28]. Copyright 2020, Wiley–VCH.

tionally, the bilayer PD was successfully fabricated on flexible PI substrates, which can be bent.

Flexible PDs based on all-inorganic perovskite (CsPbBr$_3$) QDs have been reported with improved performances and self-powered operation [28]. To enhance the optical and electrical properties of CsPbBr$_3$ QDs, a CsBr/KBr precursor was utilized during QD synthesis, as shown in Figure 2a. Under ambient conditions, the CsBr/KBr precursor solution is added to a mixture of Cs$^+$ and PbBr^2 precursors, hexane, and 2-propanol, and then the solution of colloidal CsPbBr$_3$ QDs is centrifuged, purified, and redispersed before being spin-coated. After thermal annealing, the QD thin film is formed. Figure 2b illustrates a transmission electron microscopic (TEM) image of the synthesized colloidal CsPbBr$_3$ QDs assisted by CsBr/KBr with a fraction of 0.025, which exhibit clear lattice fringes. Also, the surface quality was much improved compared to the pristine perovskite film (Figure 2c), which leads to improved carrier transport and low leakage current. The structure and photographs of flexible PDs based on CsPbBr$_3$ QDs are shown in Figure 2d. Due to the modified CsPbBr$_3$ QDs by the CsBr/KBr precursor injection, the flexible PD fabricated on the PET substrate presents enhanced performances, including a high responsivity of 10.1 AW^{-1} and detectivity of approximately 10^{14} Jones in a

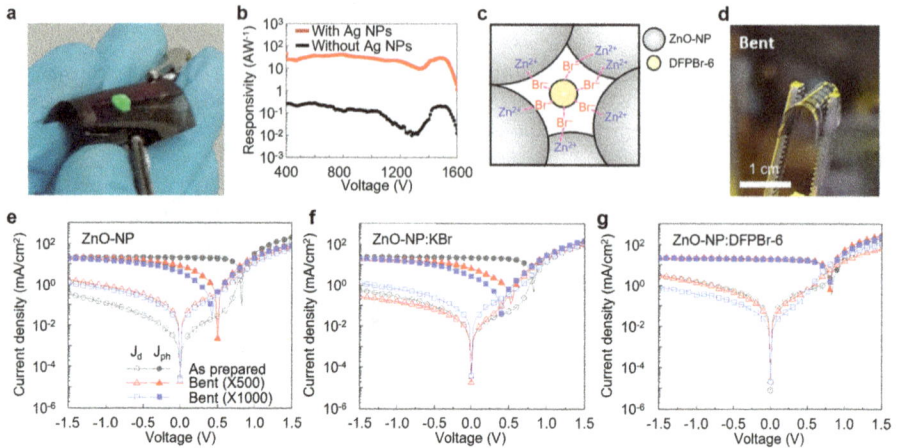

Figure 3: (Continued) 0-D material based flexible/stretchable PDs. (a) Photograph of operating flexible PD array (b) Responsivity of PbS PDs with and without Ag NPs. (a) and (b) Reproduced with permission from Zhou et al., Nat. Electron. 3, 251-258 (2020) [32]. Copyright 2020, Nature Publishing Group. (c) Schematic illustration of multivalent interactions between ZnO-NP and DFPBr-6. (d) Photograph of flexible ZnO NP: DFPBr-6 film based PDs. (e)–(g) Performance test after multiple bending cycles of (e) ZnO-NP PD, (f) ZnO-NP:KBr, and (g) ZnO-NP:DFPBr-6. (c)–(g) Reproduced with permission from Byeon et al., ACS Appl. Mater. Interfaces. (2023) [26]. Copyright 2023 American Chemical Society.

self-powered operation mode. Moreover, remarkable bending stability was observed, as shown in Figure 2e.

An upconversion photodetector was developed based on colloidal ZnO and CdSe/ZnS QDs for infrared light harvesting and visible-light emission, respectively (Figure 3a) [32]. In particular, the photodetecting layer was optimized by utilizing Ag NPs to enhance carrier tunneling, which led to a high photocurrent gain and lower dark current. The Ag NPs within the ZnO electron transport layer (ETL) facilitate electron accumulation, resulting in the bending of the valence band of the ZnO. This results in improved responsivity of the PbS PDs with a film of Ag NPs and ZnO (Figure 3b). Additionally, the fabricated PDs achieve high detectivity and rapid response time and integration with flexible substrates.

Recently, the mechanical flexibility of the ZnO NP thin film used as an ETL in flexible organic PDs were enhanced by using the multivalent interaction between ZnO NPs and fluorene pyridinium bromide derivative (DFPBr-6), one of the multicharged conjugated electrolytes [26]. The flexibility improvement in the ZnO NP thin film with DFPBr-6 results from the formation of Zn cations and Br anions bonds, as depicted in Figure 3c. Consequently, flexible organic PDs with the ZnO NP:DFPBr-6 film exhibit exquisite flexibility, sustaining a bending radius of 1.5 mm under tensile strains. The bending of the PD array is shown in Figure 3d, and the current density (J)-voltage (V) graph describes the bending stability in terms of photodetecting performance (Figure 3e–g). The flexible organic PDs with the ZnO NP:DFPBr-6 film exhibit both high photoresponse and mechanical stability (Figure 3g), compared to other PDs with ZnO-NP (Figure 3e) and ZnO-NP:KBr (Figure 3f) composite films. The presented strategy for optimizing mechanical flexibility of ZnO NP thin films suggests a way to develop flexible PDs with other metal oxide NP films.

2.1.2 One-dimensional (1D) materials-based flexible and stretchable photodetectors

Nanomaterials such as NPs, QDs, NWs, nanotubes (NTs), nanofibers, and NMs are considered powerful candidates for various optoelectronic applications including lasers [33], light-emitting diodes [34, 35], photodetectors [36], and solar cells [37], owing to their peculiar properties compared to bulk materials. Among them, 1D materials such as NWs and NTs exhibit superior optical and electrical properties, stemming from their intrinsic high surface-to-volume ratio and high photosensitivity. Additionally, 1D nanostructures demonstrate excellent mechanical flexibility due to their low physical dimensions. The fabrication methods of 1D nanomaterials can be divided into two approaches: top–down and bottom–up methods. The top–down approach involves shaping the bulk materials (e. g., Si, Ge, or GaAs) into thin nanoscale 1D structures. Initially, bulk substrates are patterned with a mask by high-resolution lithographic processes, which define geometrical features including shapes, positions, and dimensions, and then, unpatterned materials are removed during the dry- or wet-etching process [38, 39]. On the other hand, the bottom-up approach is a synthetic technique that uses seed nanoparticle deposition as a metallic catalyst to facilitate growth in a uniaxial direction. Various growth techniques such as vapor–liquid–solid (VLS) and vapor–solid–solid (VSS) methods are employed [39].

The 1D materials such as ZnO [40], ZnSe [41], CdS [42], Sb_2Se_3 [43], InGaSb [44], and $MAPbI_3$ [45] nanostructures, have been utilized as active materials for PDs. Tao et al. demonstrated flexible PDs based on ZrS_3 nanobelts to produce visible light using an adhesive-tape transfer method [46]. The nanobelt film can be integrated onto flexible substrates, including polypropylene (PP) film and printing paper, and the fabricated PDs exhibit great performance in terms of spectral sensitivity and photoresponse across a wide range from the visible to NIR. Another flexible PDs were developed based on ultra-thin Sb_2Se_3 NWs with controlled synthetic methods [43]. During the reaction of triphenylantimony with dibenzildiselenide in ethanol at a modulated temperature, oleylamine and polyvinylpyrrolidone (PVP) were utilized as surfactants. The diameters of the fabricated Sb_2Se_3 NWs range from 10 to 20 nm with lengths up to 30 μm. By adopting the Sb_2Se_3 NW film as an active layer, flexible PDs were created on PET (Figure 4a) and printing-paper substrates (Figure 4c, inset). The I-V characteristics of both PDs on PET and paper substrates exhibit good ohmic contacts between the NWs and electrodes (Figure 4b–c). The bending stability of the Sb_2Se_3 NWs PD on the PET substrate was also investigated by measuring photocurrents at various bending angles. As shown in Fig-

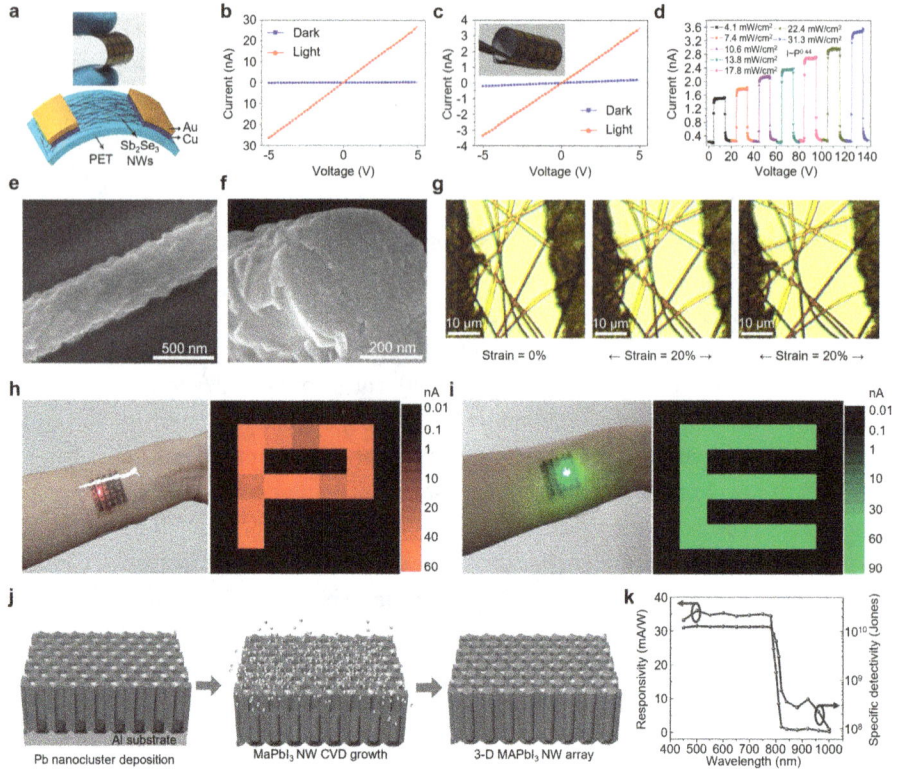

Figure 4: 1-dimensional material based flexible/stretchable PDs. (a) a photograph and illustration of flexible Sb$_2$Se$_3$ PDs on a PET substrate (b) Photoresponse of flexible PDs on a PET substrate. (c) Photoresponse of flexible PDs on a printing-paper substrate and photograph of the PDs (inset). (d) The on–off photocurrent under increased light conditions. (a)–(d) Reproduced with permission from Chen et al., Adv. Sci. 2, 1500109 (2015) [43]. Copyright 2015 John Wiley & Sons, Inc. (e)–(f) SEM images a PVP/MAPbI$_3$ nanofiber in magnified view (e) and cross-section view (f). (g) Optical images of nanofibers under external strain changes. (h)–(i) Imaging results of letter objects with red and green lights. (e)–(i) Reproduced with permission from Kim et al., ACS appl. Nano mater. 5, 1308-1316 (2022) [51]. Copyright 2022 American Chemical Society. (j) The growth process of MAPbI$_3$ NWs on PAM template. (k) Photoresponsivity and detectivity of the MAPbI$_3$ NW PDs under a 0.3 V bias. (j) and (k) Reproduced with permission from Gu et al., Adv. Mater. 28, 9713-9721 (2016) [52]. Copyright 2016, Wiley–VCH.

ure 4d, the device exhibits almost constant photocurrents regardless of bending-angle variation.

Flexible NIR PDs based on InGaSb NWs ware reported with fast response times below 20 μs [44]. The InGaSb NWs, grown by chemical vapor deposition (CVD) in two steps, are successfully arranged into a highly dense, parallel NW array on a flexible PI substrate, which enables the fabrication of flexible NIR PDs. Although the device, which lacks p–n junctions, shows a relatively high dark current compared to its silicon counterparts due to the absence of a gate layer and applied gate bias, the response times for rise and decay were measured as 13 and 16 µs, respectively, indicating a very rapid response speed. The bending stability of the flexible NIR PD was measured at various bending radii of curvature. The photoresponse results exhibit nearly constant values in all bending states, indicating not only good flexibility but also mechanical stability of the device.

Unlike conventional methods of top–down and bottom–up fabrication processes for NWs, several direct printing approaches exist, such as electrospinning [47, 48]. The advantage of electrospinning methods is their ability to fabricate various nanofibers on a large scale. In this process, a mixed solution of target materials and a polymer base is ejected through a vessel under a high voltage between the vessel tip and the collecting substrate, causing the charged solution to transform into nanofibers in a strong electric field. The physical properties, including the diameter of the collected nanofibers, are modulated by adjusting the voltage and ejection rates, and the nanofibers can be collected in either a random or uniform arrangement, depending on the addition of electrodes. Zheng et al. demonstrated a transparent, flexible, and broadband PD based on electrospun ZnO-CdO heterojunction NW arrays [49]. The hybrid ZnO-CdO NWs are fabricated by fusing ZnO and SdO precursors into individual nanofibers during the electrospinning process and are collected directly on the mica substrate for flexible and transparent PDs. The hybrid NWs exhibit broad spectral absorbance, high photoresponses, and fast response times compared to devices made only of CdO or ZnO NWs. Additionally, the PD exhibits high transparency and bending stability.

On the other hand, several approaches to enhance the performance of flexible NW PDs across wide spectral ranges based on the heterostructure of electrospun ZnO NWs and perovskite materials have been demonstrated to broaden the limited response range of ZnO NWs, which is restricted to the ultraviolet region. Zheng et al. reported flexible UV-Vis-NIR PDs using the heterostructure of electrospun ZnO NWs and PbS QD film [40], and Cao et al. demonstrated UV-Vis PDs with high performance using electrospun ZnO NWs and a perovskite film composed of CH_3NH_3I and $PbCl_2$ [50].

Stretchable PDs were demonstrated by adopting electrospun polymer and perovskite composite nanofibers [51]. The mixture of poly(vinylpyrrolidone) (PVP)/$MaPbI_3$ was electrospun to form nanofibers. The collected nanofibers exhibit unique morphologies, including the observation of nanoplates on the surface and NPs within the fibers, as shown in Figure 4e–f. In the photoluminescence analysis of a single fiber, blue-shifted double peaks were observed, stemming from the $MAPbI_3$ nanoplates and NPs. Device stretchability was explored by applying stress, and the device withstood up to 15 % strain (Figure 4g). For the wearable applications, the device was tested while attached to human skin. Under red and green light illumination, the letter objects 'P' and 'E' were clearly detected in the result images, indicating high potential for use in wearable sensing platforms (Figure 4h–i).

Meanwhile, various PDs based on vertically aligned semiconductor NWs have been introduced in recent years. Gu et al. demonstrated flexible and high-resolution image sensors based on vertically aligned lead halide perovskite NWs [52]. The growth of the

NWs is guided by a prepared nanotemplate of a free-standing porous alumina membrane (PAM) via a vapor–solid–solid-reaction (VSSR) process, using lead (Pb) metal nanoclusters as seed material. Here, the PAM not only guides the growth of the NWs but also protects them from water and oxygen molecules. The overall process of the $MAPbI_3$ NW growth in PAM is illustrated in Figure 4j. In the fabrication process, the physical dimensions of the NWs can be modulated by pre-engineering the PAM nanotemplate. With ITO top electrodes and Au metal bottom electrodes, a high-resolution flexible photodiode imager with a total of 1,024 pixels was fabricated. The spectral responsivity, as shown in Figure 4k, enables the device to detect incident optical light in the visible range, which agrees with the energy bandgap. In addition, the optical imaging results validate the potential of the device as an imager. Since the diameter of a single NW is on the nanometer scale, which is far smaller than the pixel size of conventional image sensors, the proposed design based on vertical NWs can significantly enhance the pixel resolution for next-generation image sensors.

2.1.3 2-dimensional (2D) materials-based flexible and stretchable photodetectors

Two-dimensional thin semiconducting films are an attractive candidate for flexible and stretchable PDs due to their outstanding optoelectrical and mechanical properties, such as high flexibility, transparency, and facile fabrication through exfoliation. In particular, thin films, including NMs, can easily be transferred from their donor substrates to flexible or stretchable substrates, which provides a degree of freedom in shaping the device. Recently, various flexible and stretchable PDs with high performance have been realized by thinning conventional bulk semiconducting materials, including Si [53], Ge [54], SnS_2 [55], transition metal dichalcogenides (TMDs) [56], perovskites [57], and graphene [58] thin films. Here, representative devices based on thin-film semiconductors will be introduced.

Silicon has been widely adapted to fabricate PDs due to its superior optical and electrical properties and highly matured complementary metal-oxide semiconductor (CMOS) technologies. Although its bulk form is intrinsically rigid and brittle, unsuitable for direct use in flexible devices, mechanical flexibility can be achieved through a thinning process. In this regard, several studies have been conducted using thin Si film as an active material to endow devices with not only high performance but also good mechanical flexibility. Conventional image sensors discriminate incident light of different wavelengths via optical color filter arrays, and one of the most dominant methods arranges optical filters on each pixel by dividing the sections into different colors. This method inevitably results in decreased pixel resolution, transmission loss, and an aliasing effect, which hinders high-quality imaging. To resolve these issues in traditional color imagers, flexible color PDs were demonstrated by stacking three Si membranes with different wavelength absorptions, achieved through thickness modulation [53]. Figure 5a shows a schematic of the structure and a photograph of the color PD created

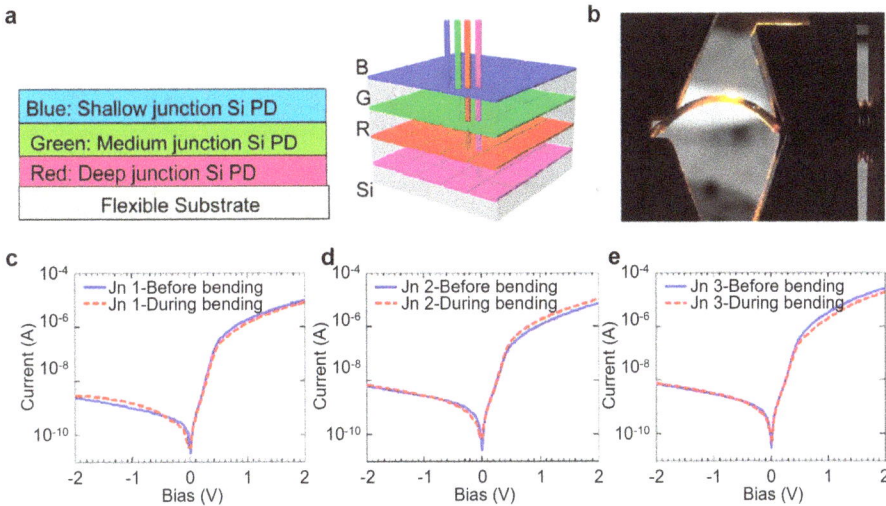

a

Blue: Shallow junction Si PD	
Green: Medium junction Si PD	
Red: Deep junction Si PD	
Flexible Substrate	

b

c

Jn 1-Before bending
Jn 1-During bending

Current (A) vs Bias (V)

d

Jn 2-Before bending
Jn 2-During bending

Current (A) vs Bias (V)

e

Jn 3-Before bending
Jn 3-During bending

Current (A) vs Bias (V)

Figure 5: 2-dimensional material based flexible/stretchable PDs. (a) Schematic illustration of multi-stacked Si based RGB PD. (b) Photograph of flexible Si PDs. (c)–(e) Bending stability tests for each junction. (a)–(e) Reproduced with permission from Menon et al., IEEE Photonics J. 7, 1–6 (2014) [53]. Copyright 2014, IEEE.

by vertically stacking three Si membranes on a flexible kapton substrate. This vertical stacking minimizes pixel resolution loss and reduces transmission loss. Additionally, the device demonstrates bending stability due to its thin film-based design on a flexible kapton substrate (Figure 5b–e). Meanwhile, Li et al. reported flexible transient PDs with Si NMs for data security purposes [59]. The Si NM-based phototransistor was created via a bottom-up thinning process (Figure 6a–b). To realize remote destruction of the device, a degradable polymer stimulated by high temperature was adopted to exert pressure on the fabricated Si phototransistors, causing them to physically self-destruct (Figure 6c).

Flexible PDs based on Ge membranes have also demonstrated for flexible and wearable optoelectronics. The flexible PDs for visible wavelengths based on Ge membranes were fabricated on a PET substrate [60]. The photodiodes are constructed to be a lateral p–i–n configuration on the Ge-on-insulator (GOI) wafer with doped regions by ion implantation and annealing process. The single-crystalline Ge membrane, with a 250-nm thickness, is placed on top of the buried oxide layer, and after the device fabrication, the PDs are removed from the base substrate by a wet etching process. Next, the dry printing technique is exploited to transfer the Ge membrane device onto the plastic substrate. The fabricated Ge devices exhibit low dark currents under several bending tests, indicating less dislocations and defects of the transferred PDs. Flexible metal–semiconductor–metal photodiodes were also developed using Ge NMs with enhanced responsivity in the NIR wavelength range [54].

In past decades, various 2D materials such as graphene and TMDs (e. g., MoS_2, WSe_2, and $MoSe_2$) have been investigated to develop flexible and high-quality PDs owing to

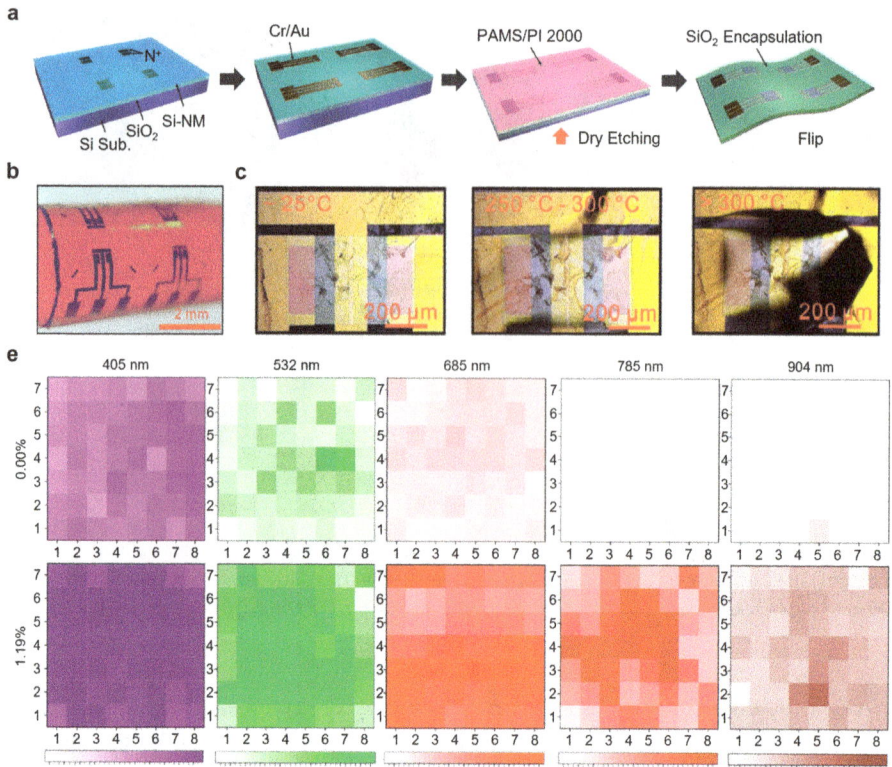

Figure 6: (Continued) 2-D material based flexible/stretchable PDs. (a) The fabrication process of Si NM phototransistors and (b) the photograph of bending state PDs. (c) Photographs of temperature-induced device destructions. (a)–(c) Reproduced with permission from Li et al., Small. 14, 1802985 (2018) [59]. Copyright Wiley-VCH GmbH, Weinheim. (d) Photoresponse changes according to the external strain. Reproduced with permission from Thai et al., ACS nano. 15, 12836–12846 (2021) [58]. Copyright 2021 American Chemical Society.

their remarkable optical, electrical, and mechanical properties. For example, flexible and transparent PDs for a broad wavelength range from UV to NIR were reported by utilizing WSe_2 films over a large area [56]. The pulsed-laser deposition (PLD) method was adopted, providing high uniformity and precise thickness control of the WSe_2 film deposition. For the transparent device, the highly crystalline WSe_2 film was deposited on the transparent PI substrate, followed by ITO deposition for transparent electrodes. The fabricated PDs show high transmittance in the visible region, approximately 72 %, while maintaining high sensitivity over a wide range of photoresponse wavelengths from 370 to 1064 nm. In addition, the device exhibits great flexibility with a small radius of curvature 5 mm, high stability in terms of photoresponse during the bending test, and withstands ambient conditions for a month.

Recently, Thai et al. demonstrated a NIR PD array by adopting strain engineering to modulate the electronic bandgap of MoS_2 [58]. MoS_2 has been adopted as an active light-absorbing material because of its high photoresponsivity owing to strong light–matter interactions, but its intrinsic characteristic of a restricted photoresponse wavelength range up to visible region has limited its use for NIR applications. To expand the response range up to the NIR region, strain engineering in the biaxial direction was adopted. Here, a metal–semiconductor–metal PD array structure was developed. The graphene membranes were utilized as the interdigitated electrodes through CVD, enhancing photocarrier separation. By applying biaxial tensile strain to the fabricated MoS_2 device via a controlled pneumatic bulge process, the PD array is able to respond to incident NIR light due to the transition of the absorption wavelength range of the MoS_2 membrane. Figure 6e shows the changes in the absorption wavelength range of the planar (Figure 6e, top column) and bulged (Figure 6e, bottom column) MoS_2 PDs. As the pressure increases, the sensitivity to NIR light is enhanced.

2.2 Mixed-dimensional materials-based flexible and stretchable photodetectors

In recent years, flexible and stretchable photodetectors based on mixed dimensional materials such as 0D-1D, 1D-2D, and 0D-2D have been demonstrated to improve the drawback of single dimensional materials, including restricted light absorption due to their intrinsic bandgap. For example, an ultraviolet 0D–1D material-based PD was fabricated using Zn_2SnO_4 NWs decorated with a ZnO QDs. Compared to the pristine Zn_2SnO_4 NW-based PDs, the device performance was remarkably improved by adding ZnO QDs to the NWs via a solvothermal method [61]. Figure 7a shows the TEM and high-resolution TEM (HRTEM) images of a ZnO QD decorated Zn_2SnO_4 NW. The advancement in device performance was induced by the decoration of ZnO QDs, which facilitates the separation of electron-hole pairs between a ZnO QD and a Zn_2SnO_4 NW, and eventually increases the electron concentration and lifetime in the Zn_2SnO_4 NW conduction channel. Figure 7b shows the photoresponsivity of the pristine Zn_2SnO_4 NW-based PD and the PD with the ZnO QD decoration. The clear peak in the ultraviolet region indicates superior wavelength selectivity of the fabricated device, and extremely higher photoresponsivity was achieved in the ZnO QD decorated PDs. Moreover, a stable photoresponse was observed under various bending conditions in terms of bending angle variation and bending cycles, as shown in Figure 7c–d.

The hybrid of 0D and 2D materials can be utilized to improve device performance, including photoresponsivity and widening the range of response wavelengths. Kolli et al. demonstrated a broadband and high responsive PD by combining SnS_2 QDs with a MoS_2 2D monolayer [62]. Here, the SnS_2 QDs, which have strong UV absorption, were adopted to overcome the responsive wavelength region of the MoS_2 monolayer, which spans from visible to NIR. To build the mixed dimensional heterostructure, the SuS_2 QDs are deposited on top of the MoS_2 monolayer using a solution-processing method (i. e., spin-coating), which is a cost-effective and low temperature process. Although the

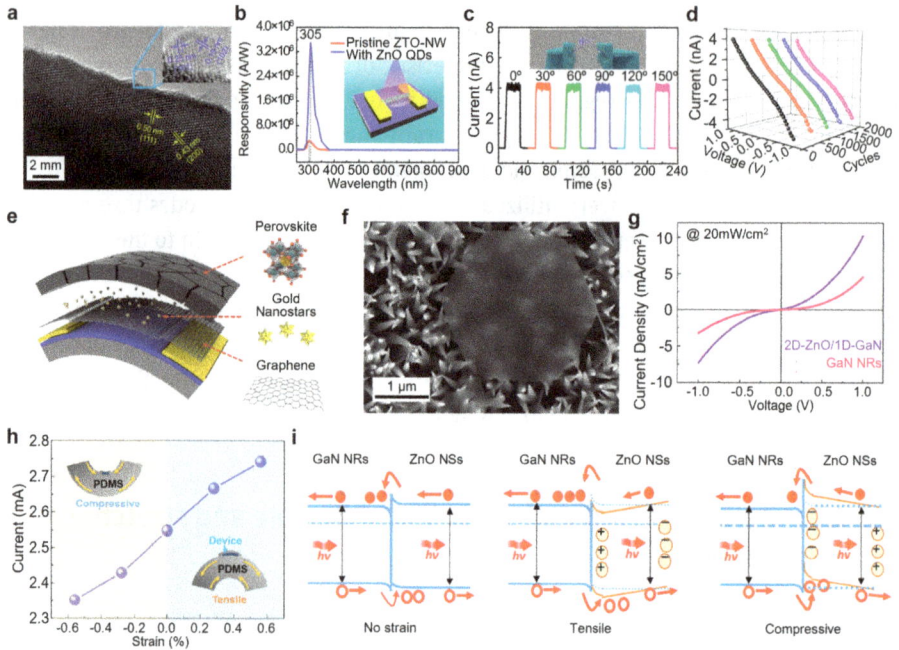

Figure 7: Mixed-dimensional material based flexible/stretchable PDs. (a) TEM images of ZnO QD decorated Zn_2SnO_4 NW. (b) The photoresponsivity of the PD based on pristine Zn_2SnO_4 NW and the ZnO QD decorated Zn_2SnO_4 NW. (c)–(d) Mechanical stability of the ZnO QD/ Zn_2SnO_4 NW PDs at various bending angles (c) and under multiple bending cycles (d). (a)–(d) Reproduced with permission from Li et al., ACS nano. 11, 4067–4076 (2017) [61]. Copyright 2017 American Chemical Society. (e) The structure of flexible and high performance hybrid PDs with perovskite film and gold nanostars. Reproduced with permission from Lee et al., NPG Asia Mater. 12, 79 (2020) [63]. Copyright 2020 Springer Nature. (f) SEM image of ZnO nanosheets on the GaN nanorods. (g) Photoresponse enhancement of the 2D ZnO/1D GaN PD. (h) Photocurrent variation under compressive and tensile strains. (i) The band diagram changes of 2D-ZnO/1D-GaN heterostructure under external strains. (f)–(i) Reproduced with permission from Lee et al., Appl. Surf. Sci. 558, 149896 (2021) [64]. Copyright 2021 Elsevier B. V.

fabricated device was based on a rigid SiO_2/Si substrate, flexible multi-dimensional PDs can be realized by constructing a heterojunction of SnS_2 on the MoS_2 layer on flexible substrates. Another hybrid PD, functionalized by gold nanostars and a perovskite layer, was reported [63] as depicted structure in Figure 7e. The gold nanostars, composed of a spherical core with sharp tips around the outer surface, were sandwiched between the top perovskite layer and bottom graphene, which exhibits a high light-trapping effect. The fabricated dual hybrid PDs on the flexible PET substrate not only presented facile and stable bendability but also high photoresponse owing to the plasmonic effect from the gold nanostars.

Flexible UV PDs based on a hybrid of 1D GaN nanorods (NRs) and 2D ZnO nanosheets were fabricated by growing the ZnO nanosheets on the GaN nanorods array via hydrothermal methods [64]. Figure 7f shows a low magnification SEM image of the grown ZnO nanosheets on GaN nanorods. The flexible 2D-ZnO/1D-GaN-based UV PD was fabricated on a PDMS substrate with bottom graphene and top carbon nanotubes (CNTs) electrodes. Compared to the device without ZnO nanosheets, the hybrid PDs exhibit enhanced photocurrent density, as shown in Figure 7g, due to the facilitated photogenerated carrier transport in the 2D-ZnO/1D-GaN heterojunction. Next, the piezo-phototronic effect was investigated by applying tensile strain to the fabricated PDs. The photoresponsivity of the device was enhanced as the tensile strain increased (Figure 7h). Under external tensile strains, piezopotential charges are induced at the heterojunction interface. Figure 7i illustrates the band diagram between the 2D-ZnO/1D-GaN heterostructure. When tensile strain is applied, the conduction band of ZnO decreases, facilitating the transfer process of generated carriers at the interface. In contrast, the photoresponsivity decreases under compressive strain because the barrier height at the interface is elevated, which hinders the carrier transport and separation.

2.3 Structural designs for flexible and stretchable photodetectors

Device flexibility and stretchability are essential for mechanical stability under temporary and occasional external variations, such as transient bending and twisting. Furthermore, PDs are adopted in various applications that require the device to retain its transformed shape over the long term, such as curved image sensors and wearable sensors. In this respect, structural designs (e. g., origami, kirigami, and island–bridge) have been intensively explored to deform 2D devices into 3D configurations.

2.3.1 Origami and kirigami structures

Origami and kirigami are types of paper arts that transform 2D paper into 3D forms using appropriate folding and cutting methods. Although origami and kirigami originate from paper-folding and paper-cutting, respectively, it is somewhat ambiguous to strictly categorize them due to the patterning process in the fabrication of flexible and stretchable PDs [65]. To date, two designs are widely used as deformable methods for curved imagers and wearable applications, owing to the facile device transformation process. For instance, hemispherical electronic eye systems were demonstrated using origami techniques in a Si-based PD array [1]. To realize a hemispherical convex PD array inspired by the retina in biological eyes, a precut single-crystalline Si PD array was fabricated on a flexible PI substrate and then was folded into a hemispherical format. Also, Choi et al. demonstrated curved PD arrays with a MoS$_2$-graphene heterostructure [2]. The 2D-based thin photodiode array was conformally transferred onto a concave hemispherical mold, owing to the predesigned structure of a truncated icosahedron, as shown in Figure 8a.

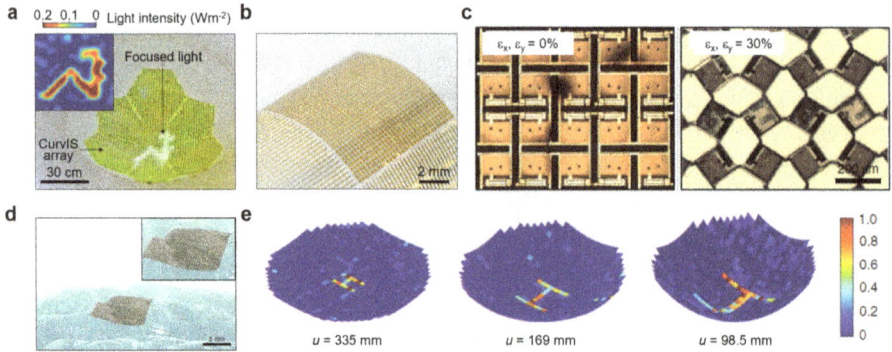

Figure 8: Structural engineering via origami/kirigami approaches. (a) A Si NM based curved image sensor with truncated icosahedron structure. Reproduced with permission from Choi et al., Nat. Commun. 8, 1664 (2017) [2]. Licensed under a Creative Commons Attribution 4.0 International License. (b) Kirigami-based flexible and stretchable PD array. (c) Photographs of unstretched and stretched kirigami-based PD array. (d) Transferred device via CAS printing on arbitrary surface. (e) Imaging results at various object distances for evaluating visual accommodation function of stretchable PDs. (b)–(e) Reproduced with permission from Rao et al., Nat. Electron. 4, 513–521 (2021) [3]. Copyright 2021, Nature Publishing Group.

Recently, a flexible and stretchable Si imager was developed that exhibits shape tunability and a high fill factor [3]. A kirigami design was adopted, featuring two perpendicular axes of reflection in one unit, and the device is composed of a repeated array of these units (Figure 8b). The fabricated PD array demonstrates a high fill factor of 78 % and withstands up to 30 % biaxial strain, as shown in Figure 8c. Also, conformal additive stamp (CAS) printing allows the device to be transferred to various curved or wavy surfaces (Figure 8d). The adaptive image sensor was constructed by transferring the kirigami Si PD array onto a shape-tunable rubber composite, which can modulate its radius of curvature using a magnetic field. With the introduction of a tunable lens, the adaptive imaging systems demonstrated the ability to continuously focus in response to changes in object distance, similar to the visual accommodation of human eyes (Figure 8e).

2.3.2 Island–bridge structures

Island–bridge designs enable to enhance the mechanical flexibility and stretchability of the PDs. In this configuration, the 'island' refers to a rigid PD region, and the 'bridge' consists of unconventionally designed electrodes, such as serpentine, watch-chain, and fractal designs, which serve as interconnections within the device. The device's flexibility and stretchability can be increased by exploiting the mechanical properties of the designed interconnects, allowing them to be bent or stretched. For example, Jung et al. demonstrated a tunable hemispherical camera system featuring silicon PDs on a thin and flexible elastomeric substrate [6]. The interconnection design of serpentine metal

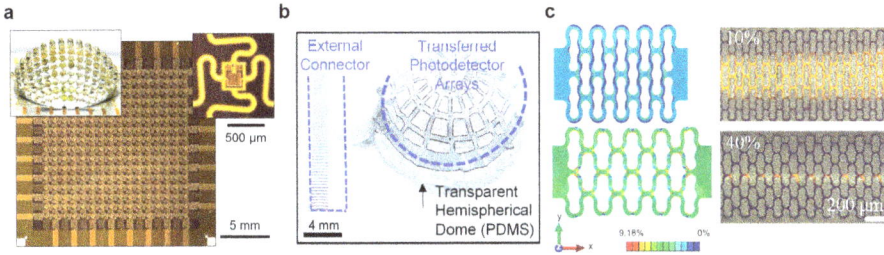

Figure 9: Structural engineering via island–bridge configuration. (a) The artificial compound eye with serpentine interconnects. Reproduced with permission from Song et al., Nature 497, 95–99 (2013) [9]. Copyright 2013 Springer Nature. (b) The spider-web inspired curved image sensor. Reproduced with permission from Lee et al., Adv. Mater. 32, 2004456 (2020) [7]. Copyright 2020, Wiley–VCH. (c) The watch-chain pattern interconnecting electrodes for wearable applications. Reproduced with permission from Li et al., Natl. Sci. Rev. 7, 849–862 (2020) [66]. Copyright 2020 China Science Publishing & Media Ltd. (Science Press).

lines provides mechanical robustness for the entire PD array. The imaging demonstration validated the adjustable zoom functions of the tunable system, successfully conducted across various radii of the device, owing to the flexible and stretchable nature of the serpentine structure. Additionally, an artificial compound eye system was demonstrated, based on the island-bridge structure with serpentine interconnects as shown in Figure 9a. Each PD is matched with a microlens, mimicking the ommatidium of an insect's eye, an optical unit, and the serpentine interconnection allows the device to be deformed into a convex configuration [9].

Meanwhile, a hemispherical PD array was fabricated based on a spider-web design with serpentine interconnects [7], as shown in Figure 9b. The fractal web design, inspired by a spiderweb, consists of a repeating fractal pattern that exhibits efficient distribution of external stress, high extensibility under stretched conditions, and great tolerance to minor cuts. The fabricated device was successfully transformed into a hemispherical form and showed high resilience even under large stress, up to 150 MPa, tolerating repetitive deformation from planer to curved or reversed formats.

A wearable health monitoring system for noninvasive blood pressure measurement was reported [66]. Including the PD array, all interconnection metal lines were designed with a watch-chain structure, which provides the device with stretchability and low impedance. The stability was tested by measuring impedance under external stress. The enhanced mechanical stability through the watch-chain structure was observed, with unchanged impedance up to a 40 % external strain condition. Figure 9c shows the finite element analysis results of the watch-chain structured system.

3 Applications of flexible and stretchable optoelectronic systems

Light signals from outside or inside of the body contain multitudes of vital information for survival. The eye, serving as one of the most sophisticated organs, collects visual data from the outside world by discriminating objects, acquiring wavelength data from the ultraviolet to the infrared region, discerning polarization, extracting depth information, and so forth. Over the past decades, the optical camera has continued to develop as an artificial vision system due to progress in semiconductor technology. Although modern camera systems sufficiently fulfill the need to acquire superior visual data with high resolution, high image contrast, and other optically specialized functions, they still need further improvements to meet the demands for a simplified and miniaturized optical system. In this regard, various approaches to realize a curved image plane by adopting flexible and stretchable PDs have been explored for advanced camera systems. In addition, recent progress in artificial vision systems inspired by biological eyes is also based on the flexible and stretchable PDs to implement unique optical functions and device miniaturization, owing to the ability for shape deformation.

On the other hand, transmitted or reflected light signals from the body enables the extraction of plentiful biological data, ranging from static images outside/inside the body (e. g., fingerprints, veins, and bones) to dynamics (e. g., pulse rate, oxygen saturation, etc.). Since the body's surface is soft and uneven compared to the flat and rigid surfaces of inorganic substrates, a gap between the soft tissues and conventional optoelectronic devices increases noise signals, causing inaccurate biosignal collection. Additionally, the biomedical application of prostheses requires the device's softness to be compatible with other organic tissues inside the body.

3.1 Biomedical sensing systems with flexible and stretchable PDs

Detecting light transmitted or reflected from the body is one of the noninvasive diagnostic tools in modern biomedical devices. Bioimaging systems are categorized into static imaging systems that aim to capture internal or surface images of the body for diagnosis and biometric authentication, and dynamic imaging systems, that collect pulse rate, blood oxidation, and blood-flow rate. For practical static biological data collection, imaging devices for static signals require high resolution rather than fast response time. In contrast, fast response time is essential in dynamic imaging systems for accurate measurements of biosignals that change over time. In addition, recent progress in wearable electronics has accelerated the development of personal health-monitoring devices for monitoring ultraviolet light levels, blood oxidation, and pulse. Lightweight, thin, and flexible devices are required to minimize user discomfort and to accumulate long-term data.

3.1.1 X-ray imaging systems

A flexible X-ray image sensor has been developed to overcome the conventional limitations of flat detectors by reducing image vignetting, which degrades accurate diagnosis due to an uneven X-ray dose over large detecting fields. In general, X-ray imager fabrication techniques are categorized into two types: the indirect conversion type, which consists of a scintillator that absorbs X-rays and emits visible light, and an optical sensor that converts light to electric charges [67]; and the direct conversion type that directly converts X-rays to electric charges [68].

One approach in the indirect conversion type is to combine a flexible scintillator with an organic PD array [69]. A bendable scintillator composed of cesium iodide (CsI) is attached to a solution-processed bulk heterojunction (BHJ) organic PD frontplane and an IGZO thin film transistor (TFT) backplane simply by pressing methods (Figure 10a). The device, stacked with a thin CsI scintillator, shows higher spatial resolution because the spread of emitted light in the lateral direction is lower than in devices with a thicker scintillator. The curved X-ray detector, with a proper radius according to the distance from the source, ensures that all X-rays are perpendicular to the detector. This prevents degradation due to incident X-rays coming from oblique directions, and enables an efficient rotating gantry for 3D X-ray imaging (Figure 10b).

On the other hand, several direct conversion type X-ray detectors have been introduced in recent years. An all-solution-based X-ray detector for large areas was developed using a printable polycrystalline $MaPBI_3$ film as an active layer, as shown in Figure 10c [70]. The fabricated device enables X-ray imaging with a low dose (Figure 10d), but the remaining work involves creating detectors on flexible substrates. However, their printing techniques for large areas have the potential to be utilized for the flexible X-ray imaging systems. Another direct conversion X-ray detector is shown in Figure 10e [71]. Here, flexible direct-conversion X-ray detector arrays is implemented using perovskite-filled membranes (PFMs). The saturated perovskite precursor solution is infiltrated into porous nylon membranes, which contain a strong fiber backbone, during the vacuum pumping process (Figure 10f). The fabricated PFMs show superior flexibility due to the porous polymer base membranes with fiber backbone (Figure 10g), and their thickness can be increased to up to mm via hot lamination, which can sufficiently stop X-rays.

3.1.2 Biometric authentication and health monitoring systems

As one of the powerful authentication methods, sensors for biometric signals, including static images of fingerprints and veins, have been intensely explored over the past decades. To enhance the security levels in biometric authentication, flexible and stretchable PDs are essential for extracting important features from acquired images. For example, there are three levels of fingerprint features: the number of ridges (level 1), branches and endings (level 2), and pores (level 3). In this regard, the fingerprint

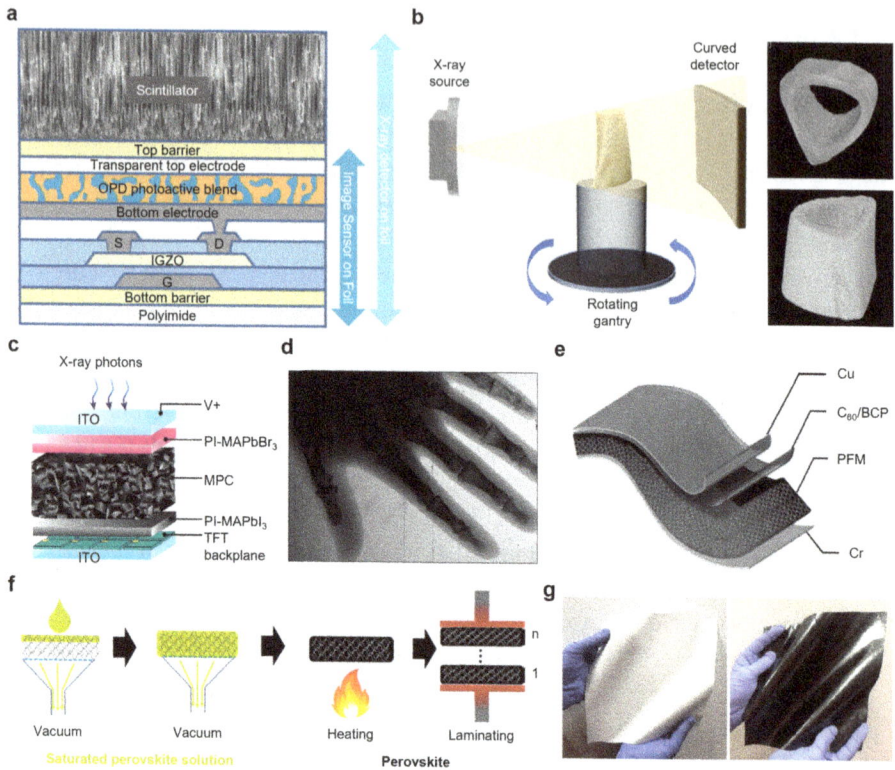

Figure 10: Flexible X-ray detectors (a) Schematic illustration of a flexible scintillator. (b) The rotating gantry for 3D X-ray imaging with curved detectors. (a)–(b) Reproduced with permission from van Breemen et al., npj Flex. Electron. 4, 22 (2020) [69]. Copyright 2020 Springer Nature. (c) The X-ray detector structure using a printable polycrystalline MaPBI$_3$ film and (d) X-ray imaging results. (e) A direct conversion X-ray detector structure. (c)–(d) Reproduced with permission from Kim et al., Nature. 550, 87–91 (2017) [70]. Copyright 2017, Nature Publishing Group. (f) Fabrication process of perovskite infiltration into porous membrane. (g) Photograph of the porous nylon membranes without (left) and with (right) the perovskite infiltration. (e)–(g) Reproduced with permission from Zhao et al., Nat. Photonics. 14, 612–617 (2020) [71]. Copyright 2020, Nature Publishing Group.

imager is required to have a resolution standard of at least 500 pixels per inch (ppi) (ISO/IEC19794-2) [72]. A flexible NIR imager composed of polycrystalline silicon TFT and organic PDs with a resolution of 508 ppi was introduced [72]. The mixing ratio of organic materials is optimized to reduce dark current while maintaining high sensitivity, and an active matrix backplane of low-temperature polycrystalline silicon (LTPS) TFT allows for high resolution and speed of the device. The conformable imager successfully captured fine details of fingerprints and veins without additional optical systems due to its flexibility, allowing direct attachment to the skin (Figure 11a–c). Breemen et al. fabricated a thin and flexible PD for fingerprints based on solution-processed metal halide perovskite PDs (Figure 11d) [73]. The dark current density was minimized to a

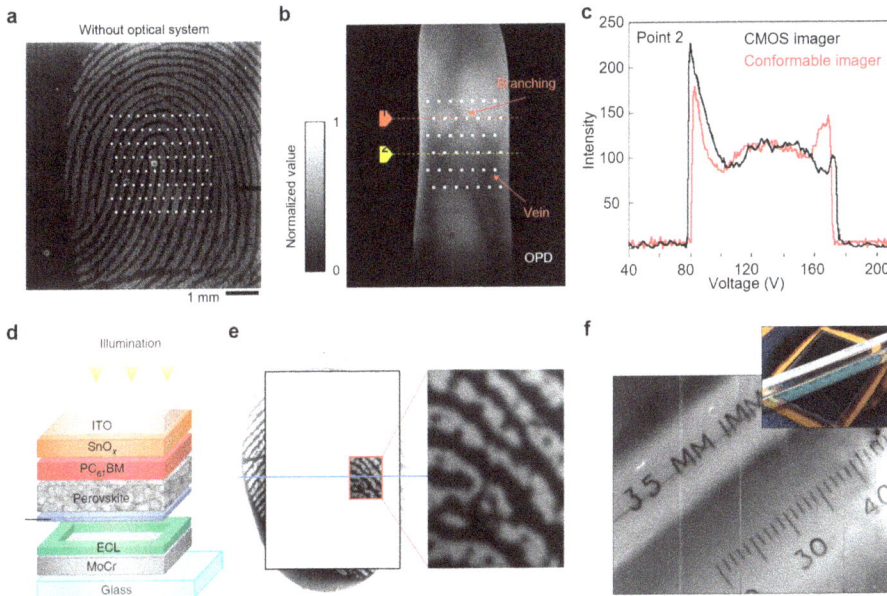

Figure 11: Flexible and stretchable imagers for bio-authentication. (a) Fingerprint imaging with conformable imager in absence of optical systems. (b) Vein imaging with conformable imager. (c) Intensity graphs of CMOS imager and conformable imager in vein images. (a)–(c) Reproduced with permission from Yokota et al., Nat. Electron. 3 113–121 (2020) [72]. Copyright 2020, Nature Publishing Group. (d) Schematic illustration of the metal halide perovskite PDs. (e) Clear fingerprint image taken by the perovskite fingerprint imager. (f) Thermometer image taken by wrapping it with a flexible sensor. (d)–(f) Reproduced with permission from van Breemen et al., Nat. Electron. 4, 818-826 (2021) [73]. Copyright 2021, Nature Publishing Group.

3×10^{-6} mA cm^{-2} by adopting a silicon nitride (SiN) edge cover layer (ECL) to cover the edges of the PD pixel, which is two orders of magnitude lower than the values in devices without ECL. The imager has a thickness of about 100 μm and a resolution of 508 ppi, as demonstrated by the clear fingerprint image (Figure 11e). Additionally, it exhibits great flexibility, enabling it to capture an image of a thermometer by wrapping around it with a radius of curvature of 0.6 cm, as shown in Figure 11f.

3.1.3 Health monitoring systems

A pulse oximetry consists of two light sources with different wavelengths and PDs for measuring light signals transmitted or reflected from the blood vessels. Two different wavelength light sources are used to discern signals from oxyhemoglobin and deoxyhemoglobin. Unlike the static biosignals, such dynamic biosignals, including photoplethysmogram (PPG) and oxygen saturation (SO$_2$), require continuous monitoring to collect accumulated data for accurate diagnosis or health maintenance. Consequently, flexible

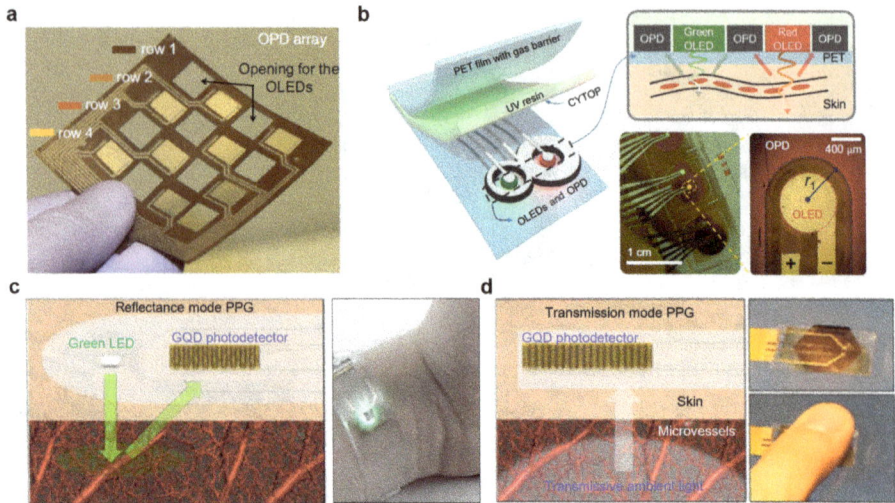

Figure 12: Flexible and stretchable PDs in biosignal sensing applications. (a) Photograph of a printed reflectance oximeter array. Reproduced with permission from Khan et al., Proc. Natl. Acad. Sci. 115, E11015-E11024 (2018) [74]. Copyright 2018 National Academy of Science. (b) Photograph of a low-energy reflective pulse oximetry patch. Reproduced with permission from Lee et al., Sci. Adv. 4, eaas9530 (2018) [75]. Copyright 2018 American Association for the Advancement of Science. (c) The reflectance mode type QD based PD and (d) the transmission mode type health patch connected to the mobile phone. (c)–(d) Reproduced with permission from Polat et al., Sci. Adv. 5, eaaw7846 (2019) [76]. Copyright 2018 American Association for the Advancement of Science.

and stretchable pulse oximeters have been investigated. Figure 12a shows a printed reflectance oximeter array based on organic materials [74]. This reflectance mode type pulse oximeter enables measurements at various locations, including the forehead, and expands the sensing area by distributing organic LEDs and PDs across a sensor array. Another reflective pulse oximetry patch with reduced power consumption was developed as shown in Figure 12b [75]. The device was fabricated as a patch-type sensor based on flexible organic LEDs and PDs, which were designed through optical simulation based on wavelength-dependent light absorption and scattering under human skin. The device configuration was ideally designed so that the organic PD is wrapped around by small organic LEDs in the shape of the number 8.

Health monitoring devices based on reflectance mode and transmission mode were demonstrated using flexible graphene sensitized with semiconducting QDs (GQD) PDs [76]. In the reflectance mode, reflected light signals are detected by GQD, as shown in Figure 12c. On the other hand, the transmission mode prototype was fabricated as a health patch connected to a mobile phone (Figure 12d). Owing to the broadband wavelength sensitivity of GQD-based PDs, transmitted light signals can be collected under ambient light conditions.

3.2 Artificial vision systems with flexible and stretchable PDs

An optical lens inherently focuses an object onto a curved image plane, called the Petzval surface, which is determined by the lens materials and geometrical curvature (Figure 13a) [24]. However, the lack of fabrication methods for curved image sensors attributes that most modern lens designs are based upon a flat image plane, which inevitably leads to multiple lens configurations, as shown in Figure 13b [77, 78]. With the increasing demand of miniaturized vision systems, several researches on curved image sensors have been proposed to reduce the number of lens elements in the imaging systems. One approach involves deforming a commercial flat CMOS sensor onto curved surfaces by applying pressure [79]. A schematic of a pneumatic forming process on the flat membrane appears in Figure 14a. Compared to deformation with fixed edges, radial tensile forces and strain energy density allow the edge die to be relieved by translation along the curved surface. Figure 14b shows a photograph of the curved status of the commercial 18-megapixel BSI CMOS sensor attached to a mold with an 18.74-mm radius of curvature. The number of lenses and aspherical surfaces can be reduced in the customized lens for curved image planes compared to the lenses for flat image sensors.

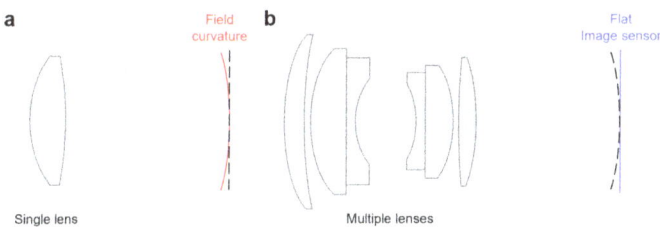

Figure 13: Schematic illustrations of the lens design for focusing on (a) the curved and (b) the planar image plane.

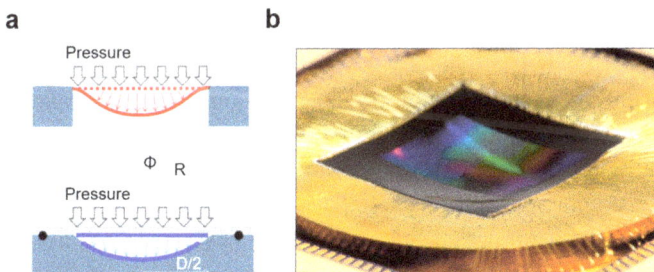

Figure 14: Schematic illustrations of (a) the pneumatic process for membrane deformation and (b) the deformed commercial image sensor. (a)–(b) Reproduced with permission from Guenter et al., Opt. Express. 25, 3010-3023 (2017) [79]. Copyright 2017 Optica Publishing Group.

Meanwhile, the natural eye structure fully exploits the advantages of the curved image plane by using a monocentric lens and a curvilinear retina, along with their unique optical properties. For instance, the human eye exhibits simple configuration of a single ball lens and curved retina, but it provides high-resolution imaging. Direct growth of perovskite NW array by using a vapor phase inside a curvilinear porous aluminum oxide membrane approach performs as a high density artificial human retina [81]. By connecting the liquid-metal wires to a data processing unit, the imaging ability of perovskite NW PD array is demonstrated, acquiring clear letter image of 'A'. Recent research in bioinspired artificial vision systems is focused onto realize both optical structures (e. g., lens, iris, and retina) and specialized imaging functions such as wide field of view, amphibious vision, polarization vision, and high sensitivity to motion. Fish eye enables to focus objects in a wide field of view with minimized optical aberrations (i. e., distortion), to conduct visual accommodation and to have a deep depth of field [8]. The ocular structure of the fish consists of a ball lens with gradient refractive index in a parabolic profile and a hemispherical retina compensate optical aberrations, providing clear panoramic images as shown in Figure 15a. Retractor and protractor muscle enable facilitation of visual accommodations by adjusting the distance between the lens and the retina. For the development of artificial fish-eye camera, half-ball lens and shell lens with different refractive indices, respectively, are integrated to focus on the hemispherical image plane. The aforementioned island–bridge structure design with serpentine interconnects allows to arrange silicon photodiode and blocking diode units along the hemispheric surface as shown in Figure 15b. Besides, nanorod texturing and surface passivation on the PD enhance both photo-absorption and light sensitivity. Another aquatic animal eye structure, the cuttlefish, is fabricated by the island–bridge design to implement a cylindrical silicon PD array (c-Si-PDA) (Figure 15c) [80]. High pixel-density distribution of specific area realizes a high-density imaging in main visual fields of the cuttlefish (Figure 15d), and flexible carbon nanotube polarization film transferred on the top of the c-Si-PDA filters out polarized light for polarization vision.

The island–bridge structure design can also be utilized to realize artificial compound eye systems. The compound eye is composed of a microlens array and photoreceptor cells matched with the individual microlenses, a structure mostly found in arthropod eyes. By transforming the flexible PD array into a convex shape, artificial compound eye systems provide a wide field of view, while achieving high sensitivity to motion [82]. The artificial compound vision inspired by the fiddler crab is shown in Figure 15e. The island–bridge design of silicon PDs and entire comb-shaped structure provide sufficient flexibility for the device to have a wide field of view of $300° \times 160°$ (H × V) (Figure 15f). Also, amphibious imaging is implemented in both air and water conditions by adopting a flat graded index microlens array and a parylene C coating for encapsulation.

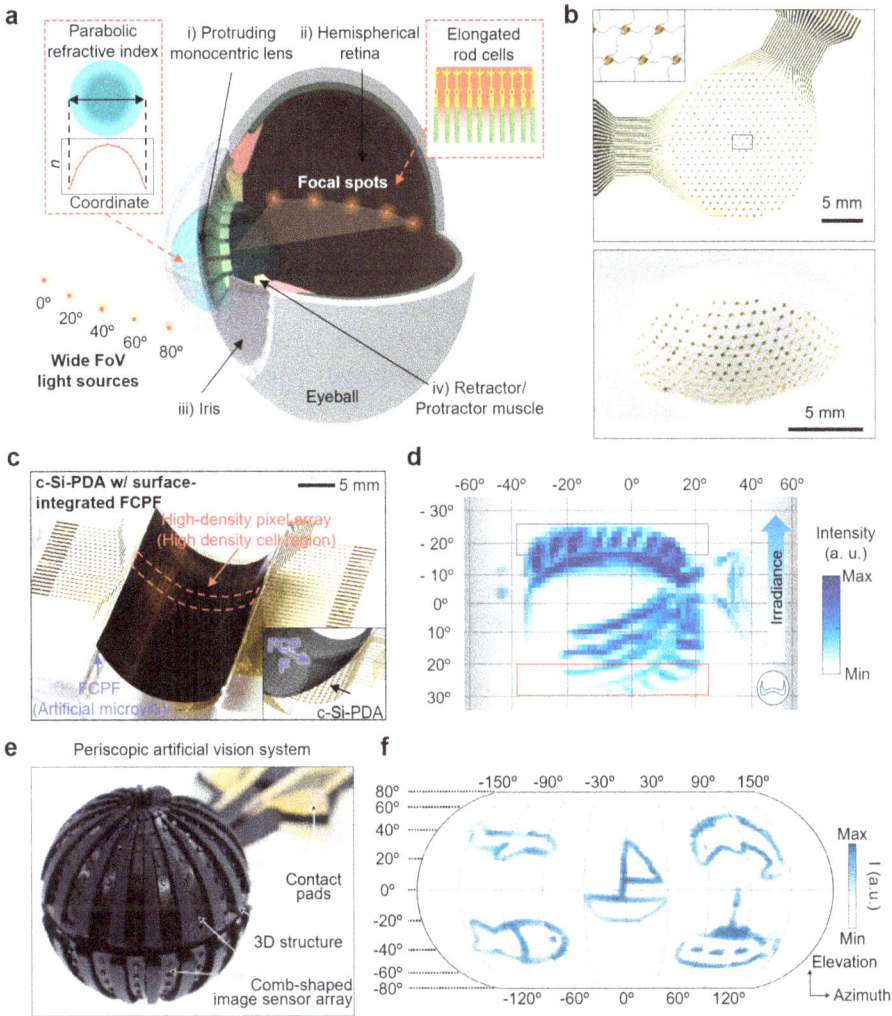

Figure 15: Structure designed, flexible and stretchable PD array for artificial vision systems. (a) The ocular structure of the fish eye. (b) Photographs of curved PD array based on the serpentine interconnects. (a)–(b) Reproduced with permission from Kim et al., Nat. Electron. 3, 546–553 (2020) [8]. Copyright 2020, Nature Publishing Group. (c) The cylindrical image sensor with pixel-density distribution and flexible polarization filter, inspired by the cuttlefish. (d) Imaging results of the cuttlefish vision systems with high-resolution region of interest (ROI) and balanced image brightness. (c)–(d) Reproduced with permission from Kim et al., Sci. Robot. 8, eade4698 (2023) [80]. Copyright 2023 American Association for the Advancement of Science. (e) Photograph of an amphibious vision system and (f) panoramic imaging results. (e)–(f) Reproduced with permission from Lee et al., Nat. Electron. 5, 452–459 (2022) [82]. Copyright 2022, Nature Publishing Group.

4 Conclusion

The technical approaches introduced in this chapter represent significant advances in flexible and stretchable PDs for next-generation optoelectronics through novel materials in nanodimensions, hybrids, and structural designs. As an essential imaging component, implementing the flexible and stretchable PD arrays enables size reduction of imaging devices beyond the limitations that conventional planar image sensors encounter. Nanodimensional material-based PDs offer a facile way to implement photodetecting systems with high performance and great flexibility, owing to their intrinsic nature of strong light–matter interactions, tunable bandgaps, and robust stability. From 0D, 1D, and 2D to mixed forms, the nanodimensional approach enhances compatibility of PDs with other optoelectronics. On the other hand, structural engineering techniques such as ultrathinning, origami/kirigami, and island-bridge have led to remarkable progress in artificial imaging systems. Artificial vision is one of the foremost research fields requiring flexible and stretchable image sensors. To overcome a key challenge of field curvature, structurally designed curved image sensors provide the next step toward state-of-the-art cameras, surpassing traditional planar CMOS imagers. The flexibility and stretchability of the sensor extend to functionalized imaging systems, including visual accommodations. In addition, flexible PDs with interconnecting electrode designs such as serpentine, fractals, and watch-chain patterns can be widely utilized in wearable and health-monitoring applications.

Bibliography

[1] Zhang, K.; Jung, Y. H.; Mikael, S.; et al. Origami silicon optoelectronics for hemispherical electronic eye systems. *Nat. Commun.* **2017**, *8* (1), 1782.

[2] Choi, C.; Choi, M. K.; Liu, S.; et al. Human eye-inspired soft optoelectronic device using high-density MoS_2-graphene curved image sensor array. *Nat. Commun.* **2017**, *8* (1), 1664.

[3] Rao, Z.; Lu, Y.; Li, Z.; et al. Curvy, shape-adaptive imagers based on printed optoelectronic pixels with a kirigami design. *Nat. Electron.* **2021**, *4* (7), 513–521.

[4] Lin, C. H.; Tsai, D. S.; Wei, T. C.; et al. Highly deformable origami paper photodetector arrays. *ACS Nano* **2017**, *11* (10), 10230–10235.

[5] Lee, W.; Liu, Y.; Lee, Y.; et al. Two-dimensional materials in functional three-dimensional architectures with applications in photodetection and imaging. *Nat. Commun.* **2018**, *9* (1), 1417.

[6] Jung, I.; Xiao, J.; Malyarchuk, V., et al. Dynamically tunable hemispherical electronic eye camera system with adjustable zoom capability. *Proc. Natl. Acad. Sci.* **2011**, *108* (5), 1788–1793.

[7] Lee, E. K.; Baruah, R. K.; Leem, J. W.; et al. Fractal web design of a hemispherical photodetector array with organic-dye-sensitized graphene hybrid composites. *Adv. Mater.* **2020**, *32* (46), 2004456.

[8] Kim, M.; Lee, G. J.; Choi, C.; et al. An aquatic-vision-inspired camera based on a monocentric lens and a silicon nanorod photodiode array. *Nat. Electron.* **2020**, *3* (9), 546–553.

[9] Song, Y. M.; Xie, Y.; Malyarchuk, V.; et al. Digital cameras with designs inspired by the arthropod eye. *Nature* **2013**, *497* (7447), 95–99.

[10] Guo, F.; Yang, B.; Yuan, Y.; et al. A nanocomposite ultraviolet photodetector based on interfacial trap-controlled charge injection. *Nat. Nanotechnol.* **2012**, *7* (12), 798–802.

[11] Oertel, D. C.; Bawendi, M. G.; Arango, A. C.; et al. Photodetectors based on treated CdSe quantum-dot films. *Appl. Phys. Lett.* **2005**, *87* (21), 213505.

[12] He, J.; Qiao, K.; Gao, L.; et al. Synergetic effect of silver nanocrystals applied in PbS colloidal quantum dots for high-performance infrared photodetectors. *ACS Photonics* **2014**, *1* (10), 936–943.

[13] Lee, W. Y.; Ha, S.; Lee, H.; et al. High-detectivity flexible near-infrared photodetector based on chalcogenide Ag$_2$Se nanoparticles. *Adv. Opt. Mater.* **2019**, *7* (22), 1900812.

[14] Shen, K.; Li, X.; Xu, H.; et al. Enhanced performance of ZnO nanoparticle decorated all-inorganic CsPbBr$_3$ quantum dot photodetectors. *J. Mater. Chem. A* **2019**, *7* (11), 6134–6142.

[15] Wu, D.; Zhou, H.; Song, Z.; et al. Welding perovskite nanowires for stable, sensitive, flexible photodetectors. *ACS Nano* **2020**, *14* (3), 2777–2787.

[16] Wang, M.; Tian, W.; Cao, F.; et al. Flexible and self-powered lateral photodetector based on inorganic perovskite CsPbI$_3$–CsPbBr$_3$ heterojunction nanowire array. *Adv. Funct. Mater.* **2020**, *30* (16), 1909771.

[17] Hossain, M.; Kumar, G. S.; Barimar Prabhava, S. N., et al. Transparent, flexible silicon nanostructured wire networks with seamless junctions for high-performance photodetector applications. *ACS Nano* **2018**, *12* (5), 4727–4735.

[18] Kim, M.; Kang, P.; Leem, J.; et al. A stretchable crumpled graphene photodetector with plasmonically enhanced photoresponsivity. *Nanoscale* **2017**, *9* (12), 4058–4065.

[19] Zheng, Y.; Cao, B.; Tang, X.; et al. Vertical 1D/2D heterojunction architectures for self-powered photodetection application: GaN nanorods grown on transition metal dichalcogenides. *ACS Nano* **2022**, *16* (2), 2798–2810.

[20] Siegmund, B.; Mischok, A.; Benduhn, J.; et al. Organic narrowband near-infrared photodetectors based on intermolecular charge-transfer absorption. *Nat. Commun.* **2017**, *8* (1), 15421.

[21] Park, S.; Fukuda, K.; Wang, M.; et al. Ultraflexible near-infrared organic photodetectors for conformal photoplethysmogram sensors. *Adv. Mater.* **2018**, *30* (34), 1802359.

[22] Zhang, Y.; Qiu, Y.; Li, X., et al. Organic single-crystalline microwire arrays toward high-performance flexible near-infrared phototransistors. *Small* **2022**, *18* (41), 2203429.

[23] Lee, G. J.; Choi, C.; Kim, D-H, et al. Bioinspired artificial eyes: Optic components, digital cameras, and visual prostheses. *Adv. Funct. Mater.* **2018**, *28* (24), 1705202.

[24] Sun, H. *Lens design: a practical guide*; CRC Press, Boca Raton, FL, USA, 2016.

[25] Cho, K. W.; Sunwoo, S. H.; Hong, Y. J.; et al. Soft bioelectronics based on nanomaterials. *Chem. Rev.* **2021**, *122* (5), 5068–5143.

[26] Byeon, H.; Kim, B.; Hwang, H.; et al. Flexible organic photodetectors with mechanically robust zinc oxide nanoparticle thin films. *ACS Appl. Mater. Interfaces* **2023**, *15* (8), 10926–10935.

[27] Ren, Z.; Sun, J.; Li, H.; et al. Bilayer PbS quantum dots for high-performance photodetectors. *Adv. Mater.* **2017**, *29* (33), 1702055.

[28] Shen, K.; Xu, H.; Li, X.; et al. Flexible and self-powered photodetector arrays based on all-inorganic CsPbBr$_3$ quantum dots. *Adv. Mater.* **2020**, *32* (22), 2000004.

[29] García de Arquer, F. P.; Armin, A.; Meredith, P.; et al. Solution-processed semiconductors for next-generation photodetectors. *Nat. Rev. Mater.* **2017**, *2* (3), 1–17.

[30] Cademartiri, L.; Montanari, E.; Calestani, G.; et al. Size-dependent extinction coefficients of PbS quantum dots. *J. Am. Chem. Soc.* **2006**, *128* (31), 10337–10346.

[31] Zhao, Y.; Li, W. PbS quantum dots band gap tuning via Eu doping. *Mater. Res. Express* **2019**, *6* (11), 115908.

[32] Zhou, W.; Shang, Y.; García de Arquer, F. P.; et al. Solution-processed upconversion photodetectors based on quantum dots. *Nat. Electron.* **2020**, *3* (5), 251–258.

[33] Eaton, S. W.; Fu, A.; Wong, A. B.; et al. Semiconductor nanowire lasers. *Nat. Rev. Mater.* **2016**, *1* (6), 1–11.

[34] Lozano, G.; Rodriguez, S. R.; Verschuuren, M. A.; et al. Metallic nanostructures for efficient LED lighting. *Light: Sci. Appl.* **2016**, *5* (6), e16080.

[35] Gao, M.; Yao, J. C.; Yan, C.; et al. Novel composite nanomaterials with superior thermal and pressure stability for potential LED applications. *J. Alloys Compd.* **2018**, *734*, 282–289.

[36] Xie, C.; Yan, F. Flexible photodetectors based on novel functional materials. *Small* **2017**, *13* (43), 1701822.

[37] Mlinar, V. Engineered nanomaterials for solar energy conversion. *Nanotechnology* **2013**, *24* (4), 042001.

[38] Song, H. S.; Lee, G. J.; Yoo, D. E.; et al. Reflective color filter with precise control of the color coordinate achieved by stacking silicon nanowire arrays onto ultrathin optical coatings. *Sci. Rep.* **2019**, *9* (1), 3350.

[39] Hobbs, R. G.; Petkov, N.; Holmes, J. D. Semiconductor nanowire fabrication by bottom-up and top-down paradigms. *Chem. Mater.* **2012**, *24* (11), 1975–1991.

[40] Zheng, Z.; Gan, L.; Zhang, J.; et al. An enhanced UV–Vis–NIR and flexible photodetector based on electrospun ZnO nanowire array/PbS quantum dots film heterostructure. *Adv. Sci.* **2017**, *4* (3), 1600316.

[41] Salfi, J.; Philipose, U.; De Sousa, C. F.; et al. Electrical properties of Ohmic contacts to ZnSe nanowires and their application to nanowire-based photodetection. *Appl. Phys. Lett.* **2006**, *89* (26), 261112.

[42] Li, L.; Lou, Z.; Shen, G. Hierarchical CdS nanowires based rigid and flexible photodetectors with ultrahigh sensitivity. *ACS Appl. Mater. Interfaces* **2015**, *7* (42), 23507–23514.

[43] Chen, G.; Wang, W.; Wang, C.; et al. Controlled synthesis of ultrathin Sb_2Se_3 nanowires and application for flexible photodetectors. *Adv. Sci.* **2015**, *2* (10), 1500109.

[44] Li, D.; Yip, S.; Li, F.; et al. Flexible near-infrared InGaSb nanowire array detectors with ultrafast photoconductive response below 20 µs. *Adv. Opt. Mater.* **2020**, *8* (22), 2001201.

[45] Wu, D.; Zhou, H.; Song, Z.; Liu, R.; Wang, H. The effect of N, N-dimethylformamide on $MAPbI_3$ nanowires for application in flexible photodetectors. *J. Mater. Chem. C* **2018**, *6* (32), 8628–8637.

[46] Tao, Y. R.; Wu, X. C.; Xiong, W. W. Flexible visible-light photodetectors with broad photoresponse based on ZrS_3 nanobelt films. *Small* **2014**, *10* (23), 4905–4911.

[47] Shui, J.; Li, J. C. Platinum nanowires produced by electrospinning. *Nano Lett.* **2009**, *9* (4), 1307–1314.

[48] Chen, H.; Wang, N.; Di, J.; et al. Nanowire-in-microtube structured core/shell fibers via multifluidic coaxial electrospinning. *Langmuir* **2010**, *26* (13), 11291–11296.

[49] Zheng, Z.; Gan, L.; Li, H.; et al. A fully transparent and flexible ultraviolet–visible photodetector based on controlled electrospun ZnO-CdO heterojunction nanofiber arrays. *Adv. Funct. Mater.* **2015**, *25* (37), 5885–5894.

[50] Cao, F.; Tian, W.; Gu, B.; et al. High-performance UV–vis photodetectors based on electrospun ZnO nanofiber-solution processed perovskite hybrid structures. *Nano Res.* **2017**, *10*, 2244–2256.

[51] Kim, H. J.; Oh, H.; Kim, T.; et al. Stretchable photodetectors based on electrospun polymer/perovskite composite nanofibers. *ACS Appl. Nano Mater.* **2022**, *5* (1), 1308–1316.

[52] Gu, L.; Tavakoli, M. M.; Zhang, D.; et al. 3D Arrays of 1024-pixel image sensors based on lead halide perovskite nanowires. *Adv. Mater.* **2016**, *28* (44), 9713–9721.

[53] Menon, L.; Yang, H.; Cho, S. J.; et al. Transferred flexible three-color silicon membrane photodetector arrays. *IEEE Photonics J.* **2014**, *7* (1), 1–6.

[54] Kim, M.; Seo, J. H.; Yu, Z.; et al. Flexible germanium nanomembrane metal-semiconductor-metal photodiodes. *Appl. Phys. Lett.* **2016**, *109* (5), 051105.

[55] Tao, Y.; Wu, X.; Wang, W.; et al. Flexible photodetector from ultraviolet to near infrared based on a SnS_2 nanosheet microsphere film. *J. Mater. Chem. C* **2015**, *3* (6), 1347–1353.

[56] Zheng, Z.; Zhang, T.; Yao, J.; et al. Flexible, transparent and ultra-broadband photodetector based on large-area WSe_2 film for wearable devices. *Nanotechnology* **2016**, *27* (22), 225501.

[57] Lee, W.; Lee, J.; Yun, H.; et al. High-resolution spin-on-patterning of perovskite thin films for a multiplexed image sensor array. *Adv. Mater.* **2017**, *29* (40), 1702902.

[58] Thai, K. Y.; Park, I.; Kim, B. J.; et al. MoS_2/graphene photodetector array with strain-modulated photoresponse up to the near-infrared regime. *ACS Nano* **2021**, *15* (8), 12836–12846.

[59] Li, G.; Song, E.; Huang, G.; et al. Flexible transient phototransistors by use of wafer-compatible transferred silicon nanomembranes. *Small* **2018**, *14* (47), 1802985.

[60] Yuan, H. C.; Shin, J.; Qin, G.; et al. Flexible photodetectors on plastic substrates by use of printing transferred single-crystal germanium membranes. *Appl. Phys. Lett.* **2009**, *94* (1), 013102.

[61] Li, L.; Gu, L.; Lou, Z.; et al. ZnO quantum dot decorated Zn_2SnO_4 nanowire heterojunction photodetectors with drastic performance enhancement and flexible ultraviolet image sensors. *ACS Nano* **2017**, *11* (4), 4067–4076.

[62] Kolli, C. S. R.; Selamneni, V.; Muñiz Martínez, B. A.; et al. Broadband, ultra-high-responsive monolayer MoS_2/SnS_2 quantum-dot-based mixed-dimensional photodetector. *ACS Appl. Mater. Interfaces* **2022**, *14* (13), 15415–15425.

[63] Lee, Y. H.; Park, S.; Won, Y.; et al. Flexible high-performance graphene hybrid photodetectors functionalized with gold nanostars and perovskites. *NPG Asia Mater.* **2020**, *12* (1), 79.

[64] Lee, D. J.; Ryu, S. R.; Kumar, G. M.; et al. Piezo-phototronic effect triggered flexible UV photodetectors based on ZnO nanosheets/GaN nanorods arrays. *Appl. Surf. Sci.* **2021**, *558*, 149896.

[65] Gao, W.; Xu, Z.; Han, X.; et al. Recent advances in curved image sensor arrays for bioinspired vision system. *Nano Today* **2022**, *42*, 101366.

[66] Li, H.; Ma, Y.; Liang, Z.; et al. Wearable skin-like optoelectronic systems with suppression of motion artifacts for cuff-less continuous blood pressure monitor. *Natl. Sci. Rev.* **2020**, *7* (5), 849–862.

[67] Douissard, P. A.; Cecilia, A.; Rochet, X.; et al. A versatile indirect detector design for hard X-ray microimaging. *J. Instrum.* **2012**, *7* (09), P09016.

[68] Yaffe, M. J.; Rowlands, J A. X-ray detectors for digital radiography. *Phys. Med. Biol.* **1997**, *42* (1), 1.

[69] van Breemen, A. J.; Simon, M.; Tousignant, O.; et al. Curved digital X-ray detectors. *npj Flex. Electron.* **2020**, *4* (1), 22.

[70] Kim, Y. C.; Kim, K. H.; Son, D. Y.; et al. Printable organometallic perovskite enables large-area, low-dose X-ray imaging. *Nature* **2017**, *550* (7674), 87–91.

[71] Zhao, J.; Zhao, L.; Deng, Y.; et al. Perovskite-filled membranes for flexible and large-area direct-conversion X-ray detector arrays. *Nat. Photonics* **2020**, *14* (10), 612–617.

[72] Yokota, T.; Nakamura, T.; Kato, H.; et al. A conformable imager for biometric authentication and vital sign measurement. *Nat. Electron.* **2020**, *3* (2), 113–121.

[73] van Breemen, A. J.; Ollearo, R.; Shanmugam, S.; et al. A thin and flexible scanner for fingerprints and documents based on metal halide perovskites. *Nat. Electron.* **2021**, *4* (11), 818–826.

[74] Khan, Y.; Han, D.; Pierre, A.; et al. A flexible organic reflectance oximeter array. *Proc. Natl. Acad. Sci. U. S. A.* **2018**, *115* (47), E11015–E11024.

[75] Lee, H.; Kim, E.; Lee, Y.; et al. Toward all-day wearable health monitoring: An ultralow-power, reflective organic pulse oximetry sensing patch. *Sci. Adv.* **2018**, *4* (11), eaas9530.

[76] Polat, E. O.; Mercier, G.; Nikitskiy, I.; et al. Flexible graphene photodetectors for wearable fitness monitoring. *Sci. Adv.* **2019**, *5* (9), eaaw7846.

[77] Kim, D. H.; Lee, G. J.; Song, Y. M. Compact zooming optical systems for panoramic and telescopic applications based on curved image sensor. *J. Opt. Microsyst.* **2022**, *2* (3), 031204.

[78] Gaschet, C.; Jahn, W.; Chambion, B.; et al. Methodology to design optical systems with curved sensors. *Appl. Opt.* **2019**, *58* (4), 973–978.

[79] Guenter, B.; Joshi, N.; Stoakley, R.; et al. Highly curved image sensors: a practical approach for improved optical performance. *Opt. Express* **2017**, *25* (12), 13010–13023.

[80] Kim, M.; Chang, S.; Kim, M.; et al. Cuttlefish eye-inspired artificial vision for high-quality imaging under uneven illumination conditions. *Sci. Robot.* **2023**, *8* (75), eade4698.

[81] Gu, L.; Poddar, S.; Lin, Y.; et al. A biomimetic eye with a hemispherical perovskite nanowire array retina. *Nature* **2020**, *581* (7808), 278–282.

[82] Lee, M.; Lee, G. J.; Jang, H. J.; et al. An amphibious artificial vision system with a panoramic visual field. *Nat. Electron.* **2022**, *5* (7), 452–459.

Hyun-Joong Chung, Rosmi Abraham, Ozge Akca Zengin,
Rayan A. M. Basodan, Shima Jalali, Jueun Lee, and Zhitong Lin

Engineering applications of materials and structures derived from stretchable electronics

Abstract: Strain gauges, electromagnetic interference (EMI) shielding, building windows, cattle trackers, and agricultural monitoring are among engineering applications that are considered as mature. This is because conventional material and device technologies have created markets that are seemingly saturated and self-content. Recently, new possibilities are emerging from the material and structure design technologies derived from stretchable electronics. In this chapter, the impacts that the emergence of stretchable electronics made to recent advances in various engineering fields are reviewed as fresh looks at the old topics.

1 Introduction

Stretchable electronics is an interdisciplinary field at the intersection of materials science, electronics, and engineering. Since their emergence in early 2000s, stretchable electronics has driven developments on 'materials that stretch' and 'structures that stretch' for many functional and structural materials that are traditionally considered as stiff and rigid [1–6]. The historical and practical significance of these materials and structures are described in Chapter 1. The mainstream applications of these material and structure innovations in academia are rather futuristic; the applications may include wearable electronics (smart clothing, fitness trackers, and medical sensors) [7], healthcare devices (skin-patch electrodes, smart wound dressing, and drug delivery systems) [8, 9], human–machine interfaces (stretchable touchscreens, smart gloves, tactile interfaces, and virtual reality/augmented reality interfaces) [10, 11], and soft robotics (soft grippers, robotic fish, and smart prosthetic limbs) [12, 13]. In this book, we covered stretchable conjugated polymeric materials (Chapter 2), stretchable liquid metal containing elastomers and soft robotics (Chapter 3) and stretchable optical devices (Chapter 4).

Even if they often seem to be overlooked, stretchable material and structure technologies have been used to tackle many longstanding problems in conventional engineering applications that are seemingly unrelated. These engineering applications are rich in history and are intertwined tightly with our society's needs to live our daily lives.

Hyun-Joong Chung, Rosmi Abraham, Ozge Akca Zengin, Rayan A. M. Basodan, Shima Jalali,
Jueun Lee, Zhitong Lin, Department of Chemical and Materials Engineering, University of Alberta,
Edmonton, Alberta T6G 1H9, Canada, e-mails: chung3@ualberta.ca, rosmi@ualberta.ca,
akcazeng@ualberta.ca, rbasodan@ualberta.ca, jueun4@ualberta.ca, zlin@ualberta.ca

https://doi.org/10.1515/9783110757286-005

In this chapter, we survey the impacts that the emergence of stretchable electronics made on recent advances in various engineering fields.

2 A case study: strain gauges in engineering applications

A strain gauge is an omnipresent device that transduces mechanical strain or deformation into an electrical signal with high precision. These strain gauges are commonly based on the piezoresistive effect of electrically active materials [14]. In other words, a strain gauge requires a material that converts minute mechanical deformations into measurable changes in electrical resistance while the material must withstand millions of load cycles with mechanical resilience. Representable engineering examples of the applications of strain gauges may include monitoring the stress distribution in an aircraft wing and high-speed train head (aerospace and rail transport industry) [15], evaluating the performance of a suspension system in a car (automotive industry) [16], and ensuring the reliability and safety of a wide range of infrastructures (civil engineering) [17, 18]. This section reviews the recent impact made by stretchable electronics on the development of next-generation strain gauges, as well as their uses in their respective engineering fields.

2.1 Essentials of strain gauges and conventional piezoresistive materials

The first generation of strain gauges were composed of discrete or point short-gauge sensors that change their resistance with respect to the pressure applied to the material that transduces; this is called piezoresistivity [14, 17]. In such sensors, an electric potential is introduced across the terminals of a conductive rod whose geometry changes with respect to the forces applied to it. The current passing through the rod exhibits a direct correlation with the applied forces. This phenomena is called piezoresistivity.

In 1856, British mathematical physicist William Thomson, also known as Lord Kelvin, demonstrated that the electrical resistance in metallic conductors changes when it is subjected to mechanical strain in his lecture "On the Electro-dynamic Qualities of Metals" [19]. This discovery sparked the initial ideas that strain can be measured electrically. German researcher Otto Schaefer pioneered the first strain gauge in 1919 [20]. His invention relied on the alteration in the vibrating frequency of strained vibrating wire in response to strain. In 1924, Burton McCullom and O. S. Peters from the U. S. Bureau of Standards conceived of a resistive strain gauge consisting of a stack of carbon discs that would change their electrical resistance when subjected to compression. Although it seems the most straightforward to measure the change of resistance across a metallic conductor while a constant direct current voltage is applied, it took 82 years to realize this idea after Lord Kelvin's demonstration; Edward E. Simmons and Arthur C. Ruge were the inventors who solved the problem of bonding agents to install the sensor in a structure [17].

A revolutionary advancement in the piezoresistive strain gauge occurred in 1954 when the piezoresistive effect was discovered in semiconductor materials of Silicon (Si) and Germanium (Ge); the burgeoning field of integrated-circuit (IC) technology played a crucial role in ushering in the first generation of commercial Si strain gauges in the late 1950s [14]. Such strain gauges are affordable, accurate, and easy to repair. However, their short gauge length restricted their use to localized strain monitoring, while monitoring at a global structural scale required different types of devices. The second constraint associated with short-gauge sensors is their limited spatial coverage within a structure, diminishing the likelihood of detecting damage. Unless damage occurs precisely where the sensors are located, the damage is unlikely to be detected [17]. The piezoresistivity-based (electrical) sensors are classified as the first generation of strain gauges.

The second-generation sensors are made of optical fibers; they offer one-dimensional distributed sensing, which enables a global structural and integrity monitoring [17]. Optical fiber technologies utilize the total internal reflection that traps light in fibers, which was demonstrated by John Tyndall in 1870. The creation of modern silica-based optical fibers as reliable guides for optical signals took nearly a century. Among many applications of the optical fibers, optical strain gauges detect changes in intensity of light passing through the fiber when strain is applied to the fiber. The optical fiber sensor is naturally capable of one-dimensional disturbed sensing with long gauge lengths. Optical sensing methods include Extrinsic Fabry–Perot Interferometry (EFPI), Michelson and Mach–Zehnder Interferometry (also known as SOFO sensors, whereas SOFO is a French abbreviation for Surveillance d'Ouvrages par Fibres Optiques), and Fiber Bragg-Grating spectrometry (FBG), as well as distributed sensing methods based on Brillouin and Rayleigh scattering [17]. Discrete optical fiber strain gauges are available in both short-gauge variants, measuring up to 10 cm in length, and long-gauge versions spanning from 25 cm to 2 m. The widely familiar plastic optical fibers with a polymethyl methacrylate (PMMA) core made their debut in the 1960s [21]. As sensors, plastic optical fibers (POFs) offer supplementary benefits, such as elevated elastic-strain limits, substantial fracture toughness, exceptional flexibility in bending, heightened sensitivity to strain, and the possibility of negative thermo-optic coefficients [22].

Distributed strain sensing utilizes an array of strain gauges to obtain spatial information of localized strains over the entire structure. For example, strain mapping is possible over a long pipeline, an airplane wing, a bridge, and a window of a high-speed train [17]. Specifically, optical fibers are capable of monitoring a 1D strain field along the entire length of beam-like structures. Laying a mesh of optical fibers can cover expansive surfaces and volumes for spatial mapping. Due to their extensive spatial coverage, distributed sensors significantly enhance the ability to detect damage and enable integrity monitoring of the entire structure (at the integrity scale) [17].

In recent years, there has been significant interest in incorporating stretchable materials in strain-gauge devices, driven by their diverse potential applications in structural health and damage monitoring, human motion detection, personal healthcare, human–machine interfaces, electronic skin, and soft robotics [23]. A notable advantage of flexible and stretchable sensor devices lies in their ability to accommodate both low and high strain levels. Traditional metal sensors are limited in their capacity to withstand strains, and inorganic semiconductor materials face limitations due to their inherent brittleness and rigidity. These traits are related to low sensitivity, limited fatigue life, environmental sensitivity, and poor biocompatibility. Integrating strain gauge in stretchable and soft platforms has the extra advantage of close integration with uneven surfaces. Various nanocomposite materials have been investigated for their potential in soft and stretchable sensors, including carbon nanotubes (CNTs), graphene, metal nanoparticles, magnetic nanoparticles, and metal nanowires. In stretchable strain gauges, these materials typically form a 3D percolation network embedded in a polymer matrix. When the composite matrix undergoes stretching, the electrical percolation between the conductive particles decreases, resulting in a reduction in electrical conductivity. Simultaneously, structural design strategies have been employed to significantly mitigate the effective strain on the conducting materials, thereby enhancing overall stretchability [24].

2.2 'Materials that stretch' that turned into next-generation strain-gauge materials

The incorporation of stretchable materials spurred the development of the next-generation piezoresistive strain gauges of distributed 2D resistive sensors and distributed stretchable optical-fiber strain gauges. Unlike conventional rigid materials that may encounter limitations at large mechanical deformation, stretchable sensor devices can be designed to detect of both low and high strain levels with irregular surfaces. In addition, these sensors can wrap uneven surfaces like a conformal skin. These qualities make them versatile in a wide range of applications, such as structural monitoring, wearable devices, human-device interfaces, and healthcare monitoring [23, 25, 26].

Distributed piezoresistive strain gauges

In numerous distributed 2D piezoresistive strain gauges, stretchable materials offer a supporting matrix. By changing the material themselves or tuning the spatial distribution of the conductive filler and the soft matrix materials, sensitivity, stretchability and response time of the strain gauges can be enhanced. The sensitivity, often quantified by a gauge factor (GF), is defined as the instant ratio of the relative change in electrical resistance to the applied mechanical stress in piezoelectric strain gauges as shown in the following equation

$$GF = \frac{(R - R_0)/R_0}{\varepsilon},$$

where R is the electrical resistance of the material, R_0 is the initial resistance, ε is the applied strain [27–30].

Highly sensitive strain gauges are useful for detecting subtle displacement in substrates, which are very beneficial in structural damage detection. The sensitivity is dependent on the piezoresistive mechanism, material choices, and the assembled structures [23]. It is also dependent on the applied strain since a nonlinear behavior is often observed with piezoresistive strain gauges [23].

When the strain gauges are based on conductive fillers, the piezoresistive mechanism occurs through the electrical tunneling between the fillers. Therefore, a subtle increase in the tunneling resistance is often effective in improving sensitivity [31, 32]. For instance, Ke et al. indicated that, by changing the mass ratio of carbon nanotubes/carbon black (CNTs/CB), they achieved a tunable electrical conductivity and strain sensitivity as shown in Figure 1a [33]. By lowering the mass ratio of CNTs/CB, the CNTs acted as the backbone of the conductive-filler phase while the CB particles bridged the neighboring CNTs to form the conductive pathways. Such an arranged conductive structure is more sensitive to changes in the mechanical strains than structures with CNTs only. When the matrix is stretched, the gaps between the CB particles and CNTs increase; hence the tunneling resistance is increased. Overall, a higher GF was achieved at low strain ranges [33]. In addition, parameters affecting the structural morphology such as process methods could also affect the GF by affecting the filler construction inside the matrix. As demonstrated by Ma et al., a strain gauge of multiwall carbon nanotubes (MWCNTs) filled with thermoplastic vulcanizate produced in one step has a higher GF of 10^3 at 100 % strain compared to that produced via a two-step synthesis [34]. This is mainly attributable to the much more disordered MWCNTs structure obtained with the one-step synthesis. Moreover, stronger interfacial bindings between the filler and matrix phases can also increase the sensitivity [33].

Another effective method to increase the sensitivity is to introduce carefully designed cracks in conductive tracks. This way, piezoresistive sensitivity can be enhanced whereas the density of cracks can determine the sensitivity and sensing ranges against the applied strain [33, 35–38]. Recently, Sun et al. developed an ultrasensitive stretchable strain gauge consisting of carbon nanotubes and silver nanowires deposited on an electrospun thermoplastic polyurethane fiber mat with prestretched microcracks shown in Figure 1b. The GF is 691 within 100 % strain, 2×10^4 from 100 % to 130 % strain, and can reach 11×10^4 with strain more than 130 % [39, 40]. Furthermore, Chen et al. demonstrated the ability to sense subtle deformation at ultra-low strain ranges with the use of crack-based strain gauges; a GF of 2×10 at strains of 0–0.3 %, 10^3 at strains of 0.3–0.5 %, and 4×10^3 at strains of 0.8–1 % are achieved [41].

In traditional applications of strain gauges, such as structural health monitoring, the required maximum level of deformation is typically less than 1 %. In wearable electronics and healthcare monitoring, however, high level of dimensional strain, as well

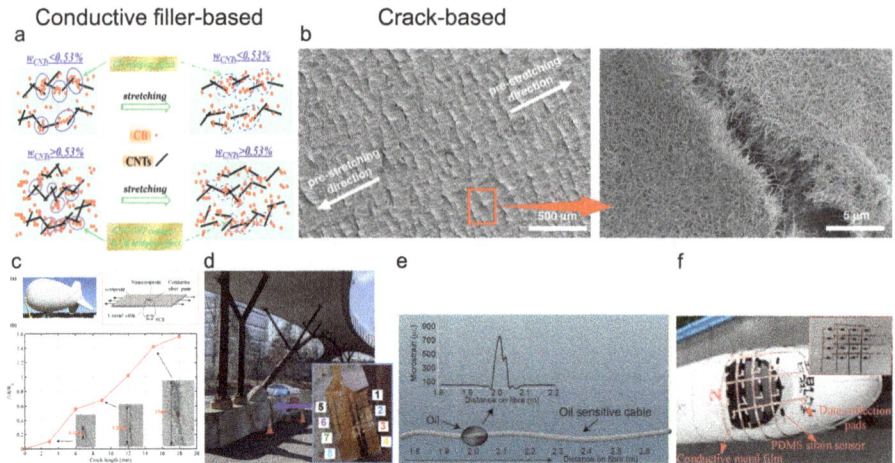

Figure 1: a) Conductive filler-based 2D distributed stretchable strain gauges, reproduced with permission from REF [33]; b) crack-based 2D distributed stretchable strain gauges, reproduced with permission from REF [40]; c) an MWCNT/CB/PDMS strain gauge used in an airship, reproduced with permission from REF [49]; d) a thin-film piezoresistive strain gauge used for bridge structure monitoring, adapted and reproduced with permission from REF [50]; e) an elastic polymer-coated optical fiber strain gauge used for oil detection along pipelines, reproduced with permission from REF [51]; f) CB/PMDS strain gauge arrays used for strain data collection in a high-speed train head, reproduced with permission from REF [52].

as a wide range of sensing, is inevitable [23]. Stretchable polymeric materials, often silicones and thermoplastic elastomers, are incorporated as an electrically inactive matrix material in this approach. Conductive filler materials are employed to provide piezoresitivity, and thus material selection and the binding between the conducting materials and matrix are often the key challenges in the field. The overall sensor productions are simple and cost-effective with high yields [28, 29, 33, 39, 42, 43].

Distributed stretchable optical-fiber strain gauges

Compared to conventional silica optical-fiber sensors, polymeric optical-fiber materials have the advantage of higher flexibility against bending, fracture toughness, and elastic strain limits [22, 44, 45]. With any minor deformation in the optical fibers' structures such as bending and stretching, the optic signal path will change accordingly [45]. The high bending flexibility and high fracture toughness are due to polymer's intrinsic mechanical properties, which are especially beneficial for use in engineering structures such as bridges and parts of buildings where a high transverse load can result in significant bending strains [22]. Another advantage is the high elastic strain limits in polymeric materials. This allows for wider applications in monitoring materials that are more elastic and flexible than conventional silica sensors. For example, PMMA optical-fiber sensors have an elastic limit of 10 % compared to 1–3 % for silica ones. In addition, PMMA optical-fiber sensors exhibit high sensitivity in both strain and temperature changes. In terms of strain sensitivity, it is characterized by the changes in the phase of a light wave travelling through an optical fiber for each unit length of elongation [22]. Silva-López et al. measured PMMA optical fiber sensors to have a GF of $131 \pm 3 \times 10^5$ rad/m, which is 15 % higher than that of bulk silica [22]. In terms of thermal sensitivity, it is characterized by the changes in the phase of a light wave travelling through an optical fiber with a unit change in temperature per unit length. For PMMA, Silva-López et al. reported it to be -212 ± 26 rad/m/K, which is almost twice that of silica (98.8 rad/m/K) in magnitude. This negative sign could be a possibility to compensate for the thermal expansion due to increased temperature in fiber sensors [22]. More accurate results can be achieved by the minimized temperature effect on the optical-fiber sensors.

Stretchability vs sensitivity

Despite the benefits of high stretchability in strain gauges, the sensitivity is often compromised by the stretchability in the strain-gauge design [17]. To have a wide range of sensing, the conductive phases must maintain their electrical percolation while still being responsive to the strain applied to the material. At the same time, these conductive phases need to be sensitive to subtle changes at very low strains [17]. Often, logic-structure design is required to achieve both high stretchability and sensitivity. For example, Amjadi et al. designed a crack-based strain gauge where graphite thin films with both short and long microcracks were coated over Ecoflex [29]. The short microcracks are responsible for the high GF of 5.2×10^2 with strains greater than 50 %, while the long microcracks have a GF of 1.1×10^4 with strains lower than 50 % [29]. A laser-patterned graphene in Ecoflex strain gauge was proposed by Tao et al. [46]. A sensitivity of 4.6×10^2 with strain up to 35 % and a GF of 2.7×10^2 with strain up to 100 % were achieved to balance the trade-off properties. Additionally, the sensing performance can be further tuned by altering the graphene mesh density [46].

Textile materials are another type of attractive stretchable platform for strain gauges especially for wearable electronics. Lee et al. designed highly sensitive multi-filament-fiber strain gauges with ultrabroad sensing ranges [47]. A layer of Ag nanoparticles was deposited on a stretchable multifilament fiber and formed an Ag-rich shell via a simple fabrication process. This structure arrangement enables the sensor to have an ultrahigh GF of 9.3×10^5 and 450 % strain in the first stretching and a GF of 659 and 200 % strain for subsequent stretching. These strain gauges were incorporated into a smart glove by sewing onto fingers. The results demonstrated successful detection of hand motions and control of a robotic hand [47]. Another example produced by Lee et al. is a stretchable and wearable strain gauge with a conductive network integrated into commercially available spandex/nylon fabrics [48]. The graphene oxide layer was deposited by a simple dip-coating method to form the conductive pathways on top of the fabrics. A high GF of 18.5 within 40.6 % strain is obtained in the coated fabric, which proved effective in detecting the bending motions of a human finger. The fabric also showed a good recovery with consistent resistance values despite the strain applied, which signifies its promising potential in wearable sensors [48]. Finally, the performance and durability of the strain gauges can improve further with many recent nanomaterial functionalities, such as self-healing, self-cleaning, biocompatibility, and optical transparency [23].

2.3 Engineering applications of the next-generation strain gauge materials

Stretchable piezoresistive strain gauges and optical-fiber sensors have been applied to large-scale engineered structures such as airplanes, bridges, pipelines, and railroads to monitor cracks and damages (Figures 1c–f). Yin et al. prepared a nanocomposite stretchable strain gauge which consists of MWCNTs and CB as the filler phase in polydimethylsiloxane (PDMS) matrix [49]. With 1.75 wt % CB and 3 wt % MWCNTs in the nanocomposite, a high GF of 12.25 is achieved up to 25 % strain with good linearity and excellent durability after 10^5 stretch–release cycles. The strain gauge detected both crack occurrence and propagation on an aerostat's surface, which evidences the sensor's potential in morphing aircraft and inflated spacecraft (Figure 1c) [49]. Piezoresistive strain gauges have also been used to monitor structural health in buildings and bridges. Aygun et al. presented a flexible array of thin-film resistive strain gauges (a GF of 2.1) applied to a pedestrian bridge (Figure 1d) [50]. The strain gauge was able to track strain changes induced by temperature variations. It was also sensitive enough to monitor early-stage damage growth of 600 µε peak strain while showing no change on crack-free surfaces in the vicinity [50].

Stretchable strain gauges can also be used to detect chemicals in large-scale engineering structures. For example, fiber-optic strain gauges have been developed for liquid hydrocarbon detection in pipelines (Figure 1e) [51]. Commercially available low-density ethylene-octene polyolefin elastomers have been employed as optical-fiber cable coatings by Totland et al. [51]. Here, strains are imparted on the fiber upon the absorption of hydrocarbons in the coating. A maximum of 700 µε is achieved with a dry fiber while a maximum of 400 µε is demonstrated with a fiber immersed in water with a response time of 1.5 minutes. This sensor design could be beneficial in detecting oil leakage along pipelines in both dry and wet environments [51].

Railroads require consistent structural health monitoring. A PDMS-based CB-filled stretchable strain gauge was produced by Cheng et al. to monitor crack formation on the curved surface of the high-speed train head [52]. An array of the strain gauges collected continuous strain data from curved surfaces of the speed train head while the train is under high-speed operation (Figure 1f) [52].

3 Other engineering applications in development

Employing materials and structures that stretch opens up unprecedented potentials of the engineering fields wherever traditional material and device technologies have been applied. In this section, a few representative examples, including electromagnetic interference (EMI) shielding, smart windows, and wearable technologies for animals, plants, and humans, are showcased. This list is, by no means, exhaustive to cover the whole spectrum of this vibrant field of technology.

3.1 Electromagnetic interference (EMI) shielding

Electronic devices, including circuit components, transportation vehicles, and medical devices, produce undesired electromagnetic waves that can result in health issues and radiation interference with the nearby electronic systems [53, 54]. EMI causes data corruption, hardware damage, slow data- transfer-speed in wireless communication, or even safety problems in more sensitive applications such as the military, as well as errors in healthcare devices [55]. As a result, shielding and blocking electromagnetic (EM) waves has become necessary to reduce the mentioned problems. There are three main mechanisms for EMI attenuation: "reflection" which occurs through the interaction of free electrons with EM waves; "absorption" which occurs through the interaction of electric or magnetic dipoles with EM radiation; and "multiple reflections" which occurs when EM waves are reflected at various surfaces or interfaces of material, especially those that include fillers. The ability to shield the EM waves can be quantified by the Shielding Effectiveness (SE) parameter which is defined as the ratio of impinging energy to the residual energy [56].

Metals have been mainly used as a material to make EMI shields due to their high electrical conductivity. However, their application has become limited considering their relatively high production cost, low corrosion resistance, heavyweight, and complexity of manufacturing into arbitrary shapes [56, 57]. Nowadays, flexible and stretchable materials are receiving increasing attention thanks to their ability to maintain their electrical conductivity and EMI shielding performance under mechanical deformation [58, 59]. These stretchable materials can be used as electromagnetic shields in daily applications such as in wearable electronics, medical devices, soft robots, and smart textiles, or even in more sensitive applications such as in the military for movable parts of weapons [60, 61].

Polymeric materials are attractive because they are easy to process, have an affordable price, and are lightweight. Their main EMI attenuation mechanism is absorption, which is useful for protecting the EM producer device itself as well as other nearby devices. There are two types of polymers that can be used as EMI shields: intrinsically conducting polymers and insulating polymer based composites [56, 62]. Li et al. studied the fabrication of a multifunctional intelligent fabric by coating a nylon fabric with metallocene (MXene) and followed by in situ polymerization of PANI as an intrinsically conducting polymer to function as a strain gauge [63]. The high electrical conductivity and specific surface area of the MXene form a conductive network that responds to deformation through real-time signal changes. This composite exhibits a shielding effectiveness value of up to 43 dB and maintains a stable response after 3,000 stretch–release cycles, reaching up to 100 % [63].

The insulating polymer-based composites needs to be compounded with conductive fillers like metal particles and carbon-based particles such as graphene or CNT to reach the required electrical conductivity for EMI shielding [62, 64]. In Feng et al.'s study, a thin film of PDMS of a thickness around 0.7 mm, filled with metal nanoparticles of Ag and Ni, was created, showing an EMI SE of 57.1 dB, as a result of the interaction of free electrons in Ag NPs' conductive network and dissipating electromagnetic wave energy [65]. Also, the composite shows a maximum tensile strain of 225 % and excellent durability through 20,000 stretch–release cycles at tensile strains ranging from 0 % to 100 % and 0 % to 150 %. Das et al. used grafted carboxylated nitrile rubber with diallyl phthalate in combination with varying contents of MWCNT to create a self-healing, stretchable nanocomposite with an EMI SE of 27.3 dB for composites containing 5 phr of filler [66]. Mani et al. demonstrated the effect of prestretching on a graphite nanoplates/polyurethane (GNP/PU) composite EMI SE with varying filler contents [67]. Their results confirmed that a 130 % prestretching of samples containing 10 % weight of GNP increased the EMI SE from 26 dB for unstretched samples to 41 dB. This behavior is attributed to the increased internal reflections and absorption achieved by forming a well-aligned GNP network. Additionally, GNP alignment reduces interfacial electrical resistance, facilitating electron movement and improving EMI shielding properties [67]. Distributing nanofillers only at the interfaces of the polymer rather than in the bulk of the material and forming an interface-reinforced segregated structure is one way to improve EMI shielding capability [56]. Based on Ma et al., results, the chemical-bond formation between PDMS microcells and the continuous PDMS interfacial phase embedded with CNT improves interface adhesion and reduces the microvoids in the CNT network, leading to an increase in EMI SE to 47.0 dB with only 2.2 vol % CNT content [68].

Hydrogels are stretchable polymeric materials in which liquid solvent components are infilled. They have the ability to change shape and adhere to various surfaces [69]. The MXene-PAA-ACC composite hydrogel exhibits a high EMI SE of 45.3 dB, utilizing an adsorption-dominated mechanism. The porous structure of the hydrogel disrupts conducting networks and increases reflection and scattering, which enhances attenuation. Additionally, water molecules inside the hydrogel can contribute to the shielding process by generating polarization loss and attenuating electromagnetic waves [70]. Zhao et al. combined polyvinyl alcohol (PVA) with a liquid metal alloy called EGaInSn and Ni microparticles to create a hydrogel with an excellent stretchability of over 800 %. The synergistic effects of conductive loss (resulting from free ions and liquid metals), dielectric loss (due to abundant dipoles in PVA and free water), and magnetic loss (stemming from the primary natural resonance of Ni) lead to a high EMI SE of 65.8 for an 8 % mass ratio of Ni [71].

Additive manufacturing is increasingly employed to fabricate complex 3D geometries in medical, aerospace, and automotive applications. Additive manufacturing is suitable for easy and rapid prototyping. Additive manufacturing is often cost-effective and leaves minimal waste of materials in manufacturing process. Therefore this method an interesting option for fabricating EMI shields [72]. Liu et al. 3D-printed a Ti_3C_2-MXene-functionalized conductive polymer hydrogel with a shielding effectiveness of 51.7 dB for a hydrogel thickness of 295 μm [73]. Shi et al. 3D printed a cellular part of polylactic acid incorporating graphene/carbon nanotube hybrid nanocomposites to achieve a 36.8 dB EMI shielding efficiency through internal multiple reflections and the conversion of electromagnetic energy to Joule heat by creating a conductive network of GNP/CNT hybrids [74].

Metamaterials are artificially fabricated materials that produce unique properties that cannot be found in nature. What enables the unique properties of the metamaterials is their specific nano-/micro-scale structures with periodicity. For EMI shielding applications, negative electrical permittivity and magnetic permeability, which is not available among naturally occurring materials, are especially useful [75]. One example of the uniqueness of metamaterial-based EMI shielding is selectively blocking EM absorbers that can block EM radiation only at certain frequencies. These types of metamaterials can be used in stealth technology and the reduction of false imaging in automotive applications. If EM waves are fully absorbed, detectors that analyze reflected components fail to detect the object [56, 76]. As another example of utilizing the concepts of metamaterials, Sakovsky et al. used the concept of mechanical metamaterial as a functional substrate for reconfigurable antennas to adapt their EM performance in response to stimuli in their environment [77]. This structure provides material with significant deformation that exceeds the elastic limit of TPU, which serves as the base material. The results demonstrate that this metamaterial-based approach enables efficient frequency reconfiguration of antennas in response to external mechanical loading. This means that, when the metamaterial is deformed, the geometry of the conductor on its surface changes, leading to a shift in the operating frequency of the antenna [77]. This concept can be use in EMI shields fabricating to manipulate and control EM waves behavior in diverse environments. In military applications, thermal camouflage is crucial for concealing objects from detection in different wavelengths of light, and metamaterial structures can be used to manipulate electromagnetic energy. Mishra et al. developed a new type of camouflage material known as flexible assembled metamaterials to provide camouflage properties in both the infrared and microwave wavelength ranges. Their samples consist of two parts, including flexible infrared emitters and flexible microwave absorbers. Their results show a 75 % reduction in contrast between the background (representing low radiative energy) and the target covered with flexible assembled metamaterials compared to a blackbody [78].

The major concepts in the stretchable EMI shields are summarized in Figure 2.

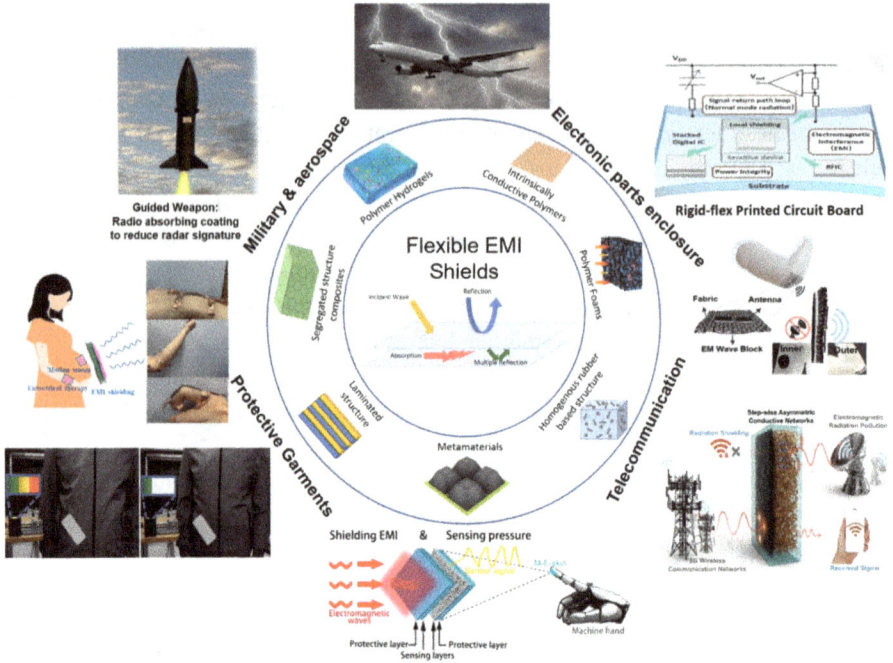

Figure 2: Schematic illustration and applications of different types of EMI shields: electronic parts enclosure, reproduced with permission from REF [79], wearable wireless communication,, reproduced with permission from REF [80], telecommunication, reproduced with permission from REF [81], shielding EMI & sensing pressure, reproduced with permission from REF [82], magnification of the signal bar of a mobile phone signal detector with the phone placed inside the jacket pocket, reproduced with permission from REF [83], and PDMA/silver nanoparticle composite strain gauges with EMI shielding ability, reproduced with permission from REF [84].

3.2 Smart windows and optical applications

Energy efficient buildings can contribute to the global effort to decrease overall energy consumption which can help mitigate the impact of humankind in climate change. Up to 50 % of all energy consumed in developed countries for building services is due to heating, ventilation, and air conditioning (HVAC) [85–89]. Cutting-edge smart windows have developed dynamically modifiable optical characteristics in response to external stimuli and have a strong potential to improve the energy efficiency [90]. It is interesting to note that the terminology of 'smart window' existed well before the term 'smart' became a defining term of modern devices like phones and televisions. Initially, the concept of smart window was proposed by researchers at Chalmers University and Lawrence Berkeley National Lab [91, 92]. In addition to chromic modulations, smart windows can have other adaptive technologies, such as moisture control, self-cleaning capabilities, dynamic coloring, and independent power production.

Smart windows offer a means to modulate indoor climate and enhance energy efficiency by tuning their colors in response to the changes in environment by the incorporation of electrochromic materials in the panes. According to environmental stimuli that triggers chromic transition, smart windows are categorized as thermochromic [93, 94], electrochromic [95, 96], photochromic [97], or humidity-chromic windows [98, 99]. Thermochromic smart windows alter their tint with respect to the environmental temperature, assisting in naturally regulating indoor temperatures. Electrochromic windows respond to the applied electric potential, making it possible to easily control their color on demand to provide on-demand privacy and light control. In photochromic window systems, color tone can be adjusted by the sunlight intensity, which is useful for reducing glare and solar heat gain. Finally, humidity-chromic windows respond to moisture levels. In humid climates, this is very useful in terms of managing light and condensation.

The temperature-sensitivity of thermochromic smart windows offer more comfort to residents without consuming any extra energy [100]. La et al. introduced a thermosensitive smart window based on polyampholyte hydrogels. It is transparent during the day and allows visible light to pass through but becomes opaque at night and blocks visible light. The phase transition occurrs above the lower critical solution temperature around 32 °C. This allows an 80 % transmittance contrast and substantial deformation capacity up to 80 % strain [101]. Among inorganic particles, vanadium dioxide (VO_2) has been extensively studied as an inorganic phase change material for energy storage and conservation applications. Its metal-insulator transition at near room temperature allows for efficient control of room temperature, which lowers energy consumption [102–104]. For example, Zhou et al. designed a novel smart window using VO_2 films or particles that regulates solar infrared radiation and scatters partial light to a solar cell for electricity generation [105]. VO_2 has a monoclinic crystalline structure below 68 °C, making it insulating and transparent to infrared radiation. However,it undergoes a tetragonal crystalline transformation above 68 °C, so, it appears metallic and reflects infrared light. Thus, the dynamic interaction of materials in thermochromic smart windows enables a reversible transition between transparent and tinted states by temperature change.

Electrochromic materials reversibly change their color through redox reactions, where ions and electrons are transferred into and out of the chromic material with respect to the applied electric potential across the window [96]. The reversible changing-color system enables indoor comfort, and privacy by regulating light and heat. Also, they are generally structured with multilayers comprising transparent conductors, electronic films, ion conductors, and ion storage films. Zhang et al. presented a controlled synthesis of plasmonic TiO_{2-x} nanocrystals with abundant oxygen vacancies, demonstrating their efficacy in dual-band electrochromic devices. The TiO_{2-x} nanocrystals exhibit strong near-infrared absorption, which enables effective control of both NIR (near infrared) and visible light with high optical modulation, fast switching speed(self-bleaching), and efficient energy recycling [106]. Cai et al. established the novel transmittance modulation of porous WO_3 films using a low-cost pulsed electrochemical deposition approach. These films have outstanding electrochromic features, including near-theoretical optical modulation (97.7 % at 633 nm), quick switching (6 and 2.7 s), high coloration efficiency (118.3 cm^2 C^{-1}), and stable cycling. The porous structure facilitates charge transfer and electrolyte penetration while also reducing WO$_3$ expansion during H+ insertion/extraction, resulting in enhanced electrochromic performance when compared to compact WO$_3$ films [107].

Photochromic materials adjust optical properties in response to ambient light conditions. Typically, these materials undergo a transition from a colorless or lightly tinted state to a darker state when activated by visible or ultraviolet (UV) light [108, 109]. Yamazaki et al. fabricated photochromic films successfully by incorporating WO$_3$ particles into a methylcellulose matrix with dispersing agents, using the solvent-casting method [110]. These biodegradable films exhibit a reversible photochromic effect, turning dark blue under UV irradiation and bleaching in the dark. The coloration rate is influenced by the structure and amounts of dispersing agents, with higher water content accelerating the process. The films demonstrate potential applications as rewritable display media by utilizing light for color changes.

Humidity-chromic (moisture-responsive) gel materials provide a self-regulating method for adjusting transparency based on humidity levels without the need for an additional energy source [111]. Inspired by adaptive color-transforming abilities of plants and animals, humidity prompts water molecules to fill microporous structures, which consequently reduces the refractive index difference to transition the opaque material to become transparent. Cai et al. developed a hydrochromic PDMS film inspired by the color changes of Diphylleia grayi, an aquatic plant [112]. To prepare the nano/microporous structure, they used LiCl in the silicone elastomer precursor. In the smart window application, the smart window gradually turns white by using a water reservoir that is connected to the gel.

Current technical challenges for smart windows include response time, reversibility, optical clarity, and undesired chemical reactions. The polymers may undergo degradation over time; exposure to environmental factors such as UV radiation, temperature fluctuations, and moisture are aggravating factors. On the other hand, enhancing response time is another essential topic to provide optimal functionality. The material must react promptly to environmental stimuli. Isolating a specific environmental stimuli factor is another issue. For example, the volume change of most hydrogel is a complex function of moisture, temperature, pH, and ion concentration. Optical clarity and reversibility between transparent and opaque states must be achieved. To achieve this, the polymerization and other chemical reactions must be optimally controlled to prevent undesired chemical reactions. Long-term environmental considerations of additive substances and degradation byproducts must be considered from the material design stage.

Polymeric smart window is an innovative advancement in building technology that tackles important problems with sustainability and energy efficiency. Exciting opportunities abound for smart window technology in the future, even if there are still issues to be resolved that require further research into new materials and technologies to improve durability, responsiveness, and compatibility with smart building systems. When the aforementioned challenges are addressed successfully, the full potential of smart windows in achieving environmentally friendly and energy-efficient building will be realized.

3.3 Stretchable sensors on animals

Throughout history, animals have played an integral role in human life, ranging from livestock to beloved pets. While animals can respond quickly to environmental changes in their own ways, wearable sensors for real-time health monitoring can be greatly beneficial because the animals often cannot understand the consequences of environmental challenges but are not able to communicate by human language. The rise of stretchable and wearable devices tailored to the soft, curved bodies of animals reflects the growing need for seamless integration.

Respiratory monitoring of large and small land animals

Understanding respiratory parameters, such as breathing rate and volume, is crucial in estimating an animal's well-being. These metrics can diagnose stress, identify cardiac and respiratory problems, and serve as early indicators of potential health issues. However, traditional monitoring methods are often ineffective especially because emotional factors from the unfamiliar device affect an animal's respiratory measuring activity, leading to unreliable clinical results. To address these challenges, the development of stretchable wearable devices becomes paramount. These devices aim to ensure accurate and continuous respiratory measurements while prioritizing animal comfort. By adapting to the soft, curved bodies of animals, wearable sensors mitigate discomfort, and thus these sensors can provide authentic insights into an animal's well-being.

Inspired by lateral-line systems in fish, which is a system of tactile sense organs, Cotur et al. introduced an air–silicone composite transducer (ASiT) as a stretchable and wireless wearable transducer [113]. ASiT resembles a rubber band when affixed to the chest or abdomen, adeptly captures volumetric changes during inhalation and exhalation. The transducer was tested on humans, dogs, and rats.

Large land animals

The increasing demand for remote health monitoring of livestock and domestic animals necessitates creating wearable sensing platforms capable of continuous data collection while the animals are in motion. For large animals like horses and cows, this demands sensors that are large in size and stretchable in form factors. In response, Chang et al. introduced a large-scale electronic textile designed for remote health monitoring for horses [114]. Electronic textiles detect physiological and electrophysiological signals on the skin, created by incorporating functional nanomaterials into fibers in specific patterns. Their dual regime spray method addresses existing challenges in e-textile production by precisely spraying conductive nanoparticles onto the fabric and coating it with a waterproof elastomer. Utilizing a programmable dual-system spray, this method achieves a customized sensor design boasting high resolution (0.9 mm in line width) and conductivity ($\geq 9{,}400\,\text{S}\,\text{cm}^{-1}$). The coating provides physical and chemical protection against use and washing. This innovative sensor design was integrated into a horse blanket and attached to a moving horse for continuous monitoring. The e-textile sensor monitored electrocardiogram signals (heart activity), electromyogram (muscle movement), and abdominal strains (respiration pattern), and demonstrated excellent accuracy in remotely monitoring the horse's health. Notably, the horse exhibited no signs of discomfort during both movement and rest, showcasing the potential of this e-textile sensor for effective and humane large-animal health monitoring.

Marine animal wearable sensors

Sea creatures are often elusive and studying their underwater behavior is not easy; expanding wearable sensor applications to marine animals can be a game changer. These sensors can provide valuable data on the physical and biological environments they traverse, especially in tracking long-distance migrations. This data is crucial to understand the ecological and conservation impacts on marine ecosystems, which are threatened by climate change and excessive fishing. However, developing wearable technology for marine animals comes with unique challenges that comes from harsh conditions in the ocean. In other words, these sensors must withstand long-term exposure to moisture, pressure, salinity conditions, and the issue of biofouling.

Nassar et al. developed "Marine Skin," a tagging system that represents a significant advancement in the field of marine research technology [115]. Crafted from polydimethylsiloxane (PDMS) and thin metal routings, this lightweight, flexible, and stretchable tag weighs less than 2.4 g in water; the design target is to make the device imperceptible to the marine animals it monitors. The device is designed to conform to various shapes and sizes of marine life, ensuring it does not interfere with natural behaviors—a critical factor in collecting accurate ecological data. The "Marine Skin" is outfitted with extensive sensor arrays that measure key environmental parameters such as pressure, temperature, and salinity. These measures are essential for understanding the dynamics of marine ecosystems and the impact of various factors on aquatic life. The tag's Bluetooth functionality facilitates wireless data transmission, offering researchers real-time, noninvasive access to vital information regarding the health of marine environments and the species that inhabit them. Field tests conducted on the swimming crab, Portunus pelagicus, showcased the "Marine Skin's" effectiveness. Despite the tag's presence, the crabs displayed mobility and behavior, possibly owing to the tag's lightweight and noninvasive design. This capability can benefit researchers aiming to gather precise data on marine ecosystems compared to the conventional bulky devices. The ultimate

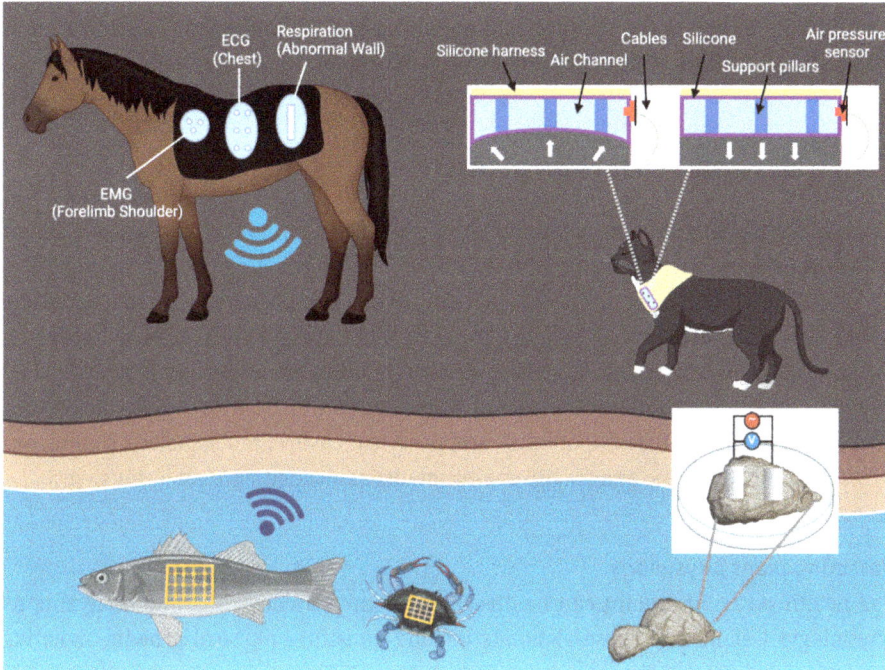

Figure 3: Wearable sensor integration in animal health monitoring. (On land) a cat is depicted wearing a respiratory sensor, while a horse is outfitted with an e-textile sensor. (In the sea) marine animals, specifically a fish and a crab are shown equipped with 'Marine Skin' wearable sensors, whereas the stress level and the environmental factors are studied on oysters.

target of "Marine Skin" is to become an innovative technology in the preservation of biodiversity.

Zhang et al. introduced a flexible bioimpedance sensor system for real-time monitoring of oyster stress levels in dynamic environments [116]. When compared to the conventional rigid counterpart, the flexible device system demonstrated 0.77 % relative error when measuring bioimpedance with phase-angle information, which offers insights into the physiological changes in oysters. Central to the study's success is the PCA-SVM model, which combines Principal Component Analysis (PCA) and Support Vector Machine (SVM) for data refinement and stress level prediction. The authors reported over 90 % accuracy in predicting stress levels. The ultimate goal of the study on the noninvasive technique is to enhance marine organism welfare and ecosystem sustainability.

In conclusion, stretchable and wearable sensors are making remarkable first steps in remotely monitoring animals. As showcased in Figure 3, the flexible and stretchable form factors of the sensors cater to the distinctive needs of various species, from nuanced respiratory measurements to large animals and marine environments. The accuracy of collected data is improved while prioritizing the comfort and well-being of the animals. As technology continues to bridge the gap between human and animal realms, the future holds promising prospects for enhanced health monitoring. This contributes to the holistic welfare of animals and deepens our understanding of their lives.

3.4 Stretchable sensors on plants

The rapidly growing human population and concurrent climate change raise a threat of food shortages. There is a critical need for water- and carbon-cycle analysis to effectively monitor plant growth. The delicate nature of plant structures, including leaves, stems, and roots, presents a challenge for conventional rigid electronics, which often result in damage to the plant. As a result, research studies are underway to develop flexible wearable sensors tailored for plants. These sensors offer several advantages such as conformal contact, biocompatibility, and precise measurements, making them more suitable for the dynamic and fragile environments associated with plant sensing.

Detecting plant physiology

The health status of a plant can be directly estimated by detecting and analyzing its physiological elements. These elements include moisture, physical growth, signaling molecules, and volatile organic compounds (VOCs). VOCs, in particular, are produced in response to external stress and serve as effective indicators of the stress factors affecting the plant.

Managing plant moisture is crucial for growth because a significant portion of a plant's weight is water. Oren et al. developed a graphene-based graphene relative-humidity sensor in the form a tape, which is designed to estimate water movement within plants [117]. The design of the sensor includes an additional air gap for efficient air exchange, whereas graphene strips on polyimide tape assess sensitivity and electrical conductivity. The sensor monitors relative-humidity changes on the leaves. Already tested with maize plants, the sensor facilitated real-time monitoring of water movement from roots to the various leaves. The technology holds promise for identifying plants with specific water-transport characteristics, crucial for crop-breeding objectives.

A plant grows by the physical expansion of stems so monitoring the growth is important. Zhang et al. introduced a sensor in a coiled-form shape that mimics the biological structure of tendrils, naming the sensor Adaptive Winding Strain [118]. This sensor is 'worn' onto tomato stems without adhesive, and transmits information about the stem's expansion and contraction to a smartphone wirelessly; they named the sensor platform the Integrated Plant Wearable System.

Li et al. introduced a wearable sensor platform for real-time detection of plant volatile organic compound (VOC) markers, focusing on a leaf-attachable chemiresistive sensor array constructed from reduced graphene oxide (rGO) functionalized with various ligands [119]. This sensor array demonstrates exceptional sensitivity and selectivity, capable of distinguishing 13 distinct plant VOCs at low to sub-ppm concentration levels with more than 97 % accuracy. The sensor's design includes a stretchable island–bridge configuration and kirigami patterns, ensuring it conforms to the dynamic changes in leaf morphology due to growth or environmental factors. Crucially, the sensor platform has been effectively applied to detect tomato late blight disease—caused by the fungus Phytophthora infestans—as early as four days post-inoculation, along with rapid detection of abiotic stresses like physical damage within one hour. This indicates the sensor's capability for continuous monitoring of VOC emissions from plant tissues. It is anticipated that the plant-wearable sensor can be utilized to enhance agricultural practices and to monitor ecosystem health.

Microclimate detection

Environmental changes, such as shifts in soil conditions, climate, and temperature, profoundly impact crop growth and physiological processes. Factors like humidity and temperature significantly influence photosynthesis. For example, higher temperatures promote more photosynthesis in general, but it may also cause closed stomata when humidity is low, leading to a decrease in photosynthesis. Especially in dense and lush environment, the microclimate that each leaf experiences differ in a localized manner. Therefore, a distributed sensor network that is directly attached to the point of interest to detect microclimate offers accurate measurements. These sensors are in close contact with the surface of a plant, making them more precise than existing long-distance

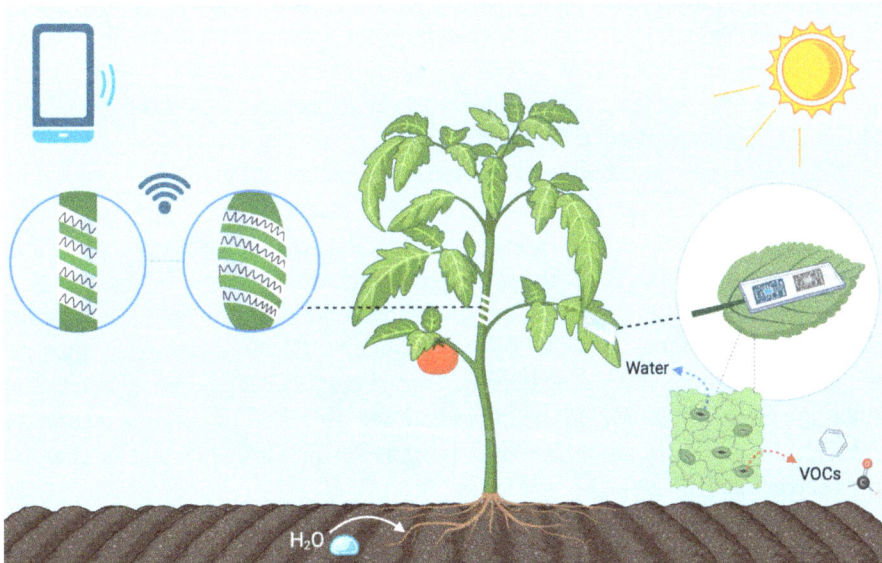

Figure 4: Conformal wearable sensor arrays for plant monitoring, showcasing the measurement of volatile organic compounds (VOCs), the monitoring of the plants' growth, and the moisture levels that the plants experience.

sensing methods. These sensors can monitor various environmental factors, including temperature, humidity, lighting, and gaseous species in the environment.

A multimodal wearable plant sensor patch attached to an abaxial side of a leaf by Lee et al. deserves a special mention [120]. This device tracked VOCs continuously while temperature and humidity conditions are simultaneously measured from the leaf and its surrounding environment, respectively. The study employed gold-coated silver nanowires (Au@AgNWs) for sensitive plant VOC detection, coupled with stability against humidity and solvent exposure.

In conclusion, various wearable plant sensors are being created to address food and environmental concerns, as summarized in Figure 4. In addition to measuring plant physiology, monitoring real-time microclimate conditions can provide valuable health assessments. Multiple sensors are integrated into a single wearable device platform to improve measurement accuracy. Remaining challenges include accuracy and cost reduction.

3.5 Engineering integrative human environments

The emergence of stretchable electronics has enabled the development of engineered interactive human environments. Personal protective equipment (PPE) transformed from providing passive protections from hazardous environments into dynamic smart systems that provide additional protections for users [121–123]. The evolution of augmented reality (AR) and virtual reality (VR) have relied heavily on stretchable electronic technology as well [11, 13, 124–127]. These engineered human environments greatly enhance safety, either directly in the case of PPE [121], or indirectly where AR/VR can be used to simulate otherwise dangerous professional training such as in aviation and firefighting [128–130]. AR/VR is also used in psychotherapy and entertainment [131, 132].

When stretchable electronics are utilized in the context of interactive human environments, they can be characterized into electronics that collect data from the user (Input), or electronics that use the processed data to influence the user (Output) [133–135]. In the application of smart PPE, the input is typically sensed danger, both due to hazardous environments as well as the health status of the user. The output is typically informing the user of the presence of dangers that are difficult to perceive or pressuring the user into evacuating dangerous circumstances [136, 137]. In the application of AR/VR, the input is typically information about the user such as position, motion, or health indicators. AR/VR output is more varied as the output's function is to alter the user's perception [127, 138].

In both example applications of smart PPE and AR/VR, stretchable electronic sensors are key. The stretchable sensors collect user and environmental data as input to be processed. Hypothermia, hyperthermia, and impact-detection sensors are useful to collect data from users in smart PPE applications. Stretchable sensors for biohazards, toxic fumes, flame, vibration, and heat/cold among others are useful for environmental input in PPE applications. Stretchable sensors for strain, acceleration, angular motion, and position are valuable input sources for AR/VR applications. Heart rate, respiration, and perspiration stretchable sensors can be utilized in both example engineered interactive human environments. Sensors embedded in the user's environment such as professional tools can also be used as input in the application of smart PPE [121, 122, 139–143].

The output in AR/VR predominantly includes the use of screens to alter the user's visual perception of reality. Rigid screens, flat or curved, have historically been used in VR headsets, or by utilizing screens pervading day-to-day life, such as smartphone screens. Flexible thin screens enabled by stretchable electronics are paving the way for the future of AR/VR. These flexible screens can be overlayed human skin or worn as contact lenses. Specialists make the distinction between "conventional" and "skin contact" technology when comparing rigid and flexible electronic technologies for engineering interactive human environments. Vision is only one sense influenced by screens. For more impactful technologies, the other senses are leveraged as well. Haptics is the stimulation the tactile sense, including motion and vibration. In AR/VR, haptics is used to stimulate the user's tactile sense to convince the user of the simulated environment, creating a more immersive experience. In the PPE application, vibration creates a sense of urgency when informing the user of a hazard. Many professionals who are required to don PPE have the tendency to overestimate their capabilities. So, creating a sense of urgency is important for self-rescue to be initiated before it is no longer possible. Hearing is yet another sense leveraged to influence perceived environments in both AR/VR and smart PPE. However, there is still more potential for flexible electronics to enhance these technologies [11, 13, 124–126, 144–152].

4 Conclusion

The progress of technology is highly nonlinear. Idea and inspirations from seemingly unrelated fields sometimes drive a major breakthrough in a long-studied field of technology. Materials science technologies that have been driving the advance of the field of stretchable electronics, such as stretchable (semi)conducting materials, origami/kirigami techniques, and mechanical metamaterials, are now providing fresh design motifs for the fields of engineering that have traditionally been considered old and established. In this chapter, recent progresses in strain- gauge technology using technologies from the stretchable electronics field is reviewed in-depth. Case examples from EMI shielding and smart window are showcased. Finally, we speculate on the impact of wearable technologies in studying animals and plants and in enhancing human perceptions.

Bibliography

[1]   Jones, J.; Lacour, S. P.; Wagner, S.; Suo, Z. Stretchable wavy metal interconnects. *J. Vac. Sci. Technol. A* **2004**, *22*, 1723–1725.
[2]   Someya, T.; Kato, Y.; Sekitani, T.; Iba, S.; Noguchi, Y.; Murase, Y.; Kawaguchi, H.; Sakurai, T. Conformable, flexible, large-area networks of pressure and thermal sensors with organic transistor active matrixes. *Proc. Natl. Acad. Sci. USA* **2005**, *102* (35), 12321–12325.
[3]   Sun, Y.; Choi, W. M.; Jiang, H.; Huang, Y. Y.; Rogers, J. A. Controlled buckling of semiconductor nanoribbons for stretchable electronics. *Nat. Nanotechnol.* **2006**, 1, 201–207.
[4]   Khang, D. Y.; Jiang, H.; Huang, Y.; Rogers, J. A. A stretchable form of single-crystal silicon for high-performance electronics on rubber substrates. *Science* **2006**, *311*, 208–212.
[5]   Loher, T.; Manessis, D.; Heinrich, R.; Schmied, B.; Vanfleteren, J.; DeBaets, J.; Ostmann, A.; Reichl, H. Stretchable electronic systems. In *Proceedings of the electronic packaging technology conference*; EPTC, 2006; pp. 271–276; 414725.
[6]   Brosteaux, D.; Axisa, F.; Gonzalez, M.; Vanfleteren, J. Design and fabrication of elastic interconnections for stretchable electronic circuits. *IEEE Electron Device Lett.* **2007**, *28* (7), 552–554.
[7]   Huang, Q.; Zhu, Y. Printing conductive nanomaterials for flexible and stretchable electronics: a review of materials, processes, and applications. *Adv. Mater. Technol.* **2019**, *4*, 1800546.
[8]   Zhu, P.; Peng, H.; PRwei, A. Y. Flexible, wearable biosensors for digital health. *Med. Novel Tech. Dev.* **2022**, *14*, 100118.
[9]   Herbert, R.; Lim, H. R.; Park, S.; Kim, J. H.; Yeo, W. H. Recent advances in printing technologies of nanomaterials for implantable wireless systems in health monitoring and diagnosis. *Adv. Healthc. Mater.* **2024**, *10*, 2100158.
[10]   Wu, H.; Yang, G.; Zhu, K.; Liu, S.; Guo, W.; Jiang, Z.; Li, Z. Materials, devices, and systems of on-skin electrodes for electrophysiological monitoring and human–machine interfaces. *Adv. Sci.* **2021**, *8*, 2001938.
[11]   Liu, S.; Ma, K.; Yang, B.; Li, H.; Tao, X. Textile Electronics for VR/AR Applications. *Adv. Funct. Mater.* **2021**, *31*, 2007254.
[12]   Shintake, J.; Cacucciolo, V.; Floreano, D.; Shea, H Soft Robotic Grippers. *Adv. Mater.* **2018**, *30*, 1707035.
[13]   Yang, T. H.; Kim, J. R.; Jin, H.; Gil, H.; Koo, J. H.; Kim, H. J. Recent Advances and Opportunities of Active Materials for Haptic Technologies in Virtual and Augmented Reality. *Adv. Funct. Mater.* **2021**, *31*, 2008831.
[14]   Fiorillo, A. S.; Critello, C. D.; Pullano, S. A. Theory, technology and applications of piezoresistive sensors: a review. *Sens. Actuators A, Phys.* **2018**, *281*, 156–175.
[15]   Cheng, X.; Bao, C.; Wang, X.; Dong, W. Stretchable strain sensor based on conductive polymer for structural health monitoring of high-speed train head. *Proc. Inst. Mech. Eng. Pt. L J. Mater. Des. Appl.* **2020**, *234* (3), 496–503.
[16]   Corvito, D.; Dindo, L.; Vendramin, M.; Menaghetti, G. Development of methods to measure the forces at the rear suspension of a race car during racetrack driving. *IOP Conf. Ser., Mater. Sci. Eng.* **2022**, *1214*, 012048.
[17]   Glisic, B. Concise historic overview of strain sensors used in the monitoring of civil structures: the first one hundred years. *Sensors* **2022**, *22*, 2397.
[18]   Loupos, K.; Damigos, Y.; Amditis, A.; Gerhard, R.; Rychkov, D.; Wirges, W.; Schulze, M.; Lenas, S. A.; Chatiandeoglou, C.; Malliou, C. M.; Tsaoussidis, V.; Brady, K.; Frankenstein, B. Structural health monitoring system for bridges based on skin-like sensor. *IOP Conf. Ser., Mater. Sci. Eng.* **2017**, *236*, 012100.
[19]   Torfs, T.; Sterken, T.; Brebels, S.; Santana, J.; van den Hoven, R.; Spiering, V.; Bertsch, N.; Trapani, D.; Zonta, D. Low power wireless sensor network for building monitoring. *IEEE Sens. J.* **2013**, *13*, 909–915.
[20]   Schaefer, O. Die Schwingende Saite als Dehnungsmesser. *Zeil. Des. Ver. Dtsch. Ing.* **1919**, *63*, 1008.

[21]   Zubia, J.; Arrue, J. Plastic optical fibers: an introduction to their technological processes and applications. *Opt. Fiber Technol.* **2001**, *7*, 101–140.

[22]   Peters, K. Polymer optical fiber sensors—a review. *Smart Mater. Struct.* **2011**, *20*, 013002.

[23]   Duan, L.; D'hooge, D. R.; Cardon, L. Recent progress on flexible and stretchable piezoresistive strain sensors: From design to application. *Prog. Mater. Sci.* **2020**, *114*, 100617.

[24]   Wu, S.; Moody, K.; Kollipara, A.; Zhu, Y. Highly sensitive, stretchable, and robust strain sensor based on crack propagation and opening. *ACS Appl. Mater. Interfaces* **2023**, *15*, 1798–1807.

[25]   Shieh, J.; Huber, J. E.; Fleck, N. A.; Ashby, M. F. The selection of sensors. *Prog. Mater. Sci.* **2001**, *46* (3–4), 461–504.

[26]   Dobie, W. B.; Isaac, P. C. G.; Gale, G. O. Electrical resistance strain gauges. *Am. J. Phys.* **1950**, *18*, 117–118.

[27]   Yang, T.; Xie, D.; Li, Z.; Zhu, H. Recent advances in wearable tactile sensors: materials, sensing mechanisms, and device performance. *Mater. Sci. Eng. Rep.* **2017**, *115*, 1–37.

[28]   Choi, D. Y.; Kim, M. H.; Oh, Y. S.; Jung, S. H.; Jung, J. H.; Sung, H. J.; Lee, H. W.; Lee, H. M. Highly stretchable, hysteresis-free ionic liquid-based strain sensor for precise human motion monitoring. *ACS Appl. Mater. Interfaces* **2017**, *9* (2), 1770–1780.

[29]   Amjadi, M.; Kyung, K. U.; Park, I.; Sitti, M. Stretchable, skin-mountable, and wearable strain sensors and their potential applications: a review. *Adv. Funct. Mater.* **2016**, *26* (11), 1678–1698.

[30]   Liu, Z.; Qi, D.; Guo, P.; Liu, Y.; Zhu, B.; Yang, H.; Liu, Y.; Li, B.; Zhang, C.; Yu, J.; Liedberg, B.; Chen, X. Thickness-gradient films for high gauge factor stretchable strain sensors. *Adv. Mater.* **2015**, *27* (40), 6230–6237.

[31]   Hu, N.; Karube, Y.; Arai, M.; Watanabe, T.; Yan, C.; Li, Y.; Liu, Y.; Fukunaga, H. Investigation on sensitivity of a polymer/carbon nanotube composite strain sensor. *Carbon* **2010**, *48* (3), 680–687.

[32]   Zare, Y.; Rhee, K. Y. Formulation of tunneling resistance between neighboring carbon nanotubes in polymer nanocomposites. *Eng. Sci. Technol. Int. J.* **2021**, *24* (3), 605–610.

[33]   Ke, K.; Pötschke, P.; Wiegand, N.; Krause, B.; Voit, B. Tuning the network structure in poly(vinylidene fluoride)/carbon nanotube nanocomposites using carbon black: toward improvements of conductivity and piezoresistive sensitivity. *ACS Appl. Mater. Interfaces* **2016**, *8* (22), 14190–14199.

[34]   Ma, L. F.; Bao, R. Y.; Dou, R.; Zheng, S. D.; Liu, Z. Y.; Zhang, R. Y.; Yang, M. B.; Yang, W. Conductive thermoplastic vulcanizates (tpvs) based on polypropylene (pp)/ethylene-propylene-diene rubber (epdm) blend: from strain sensor to highly stretchable conductor. *Compos. Sci. Technol.* **2016**, *128*, 176–184.

[35]   Zhou, J.; Yu, H.; Xu, X.; Han, F.; Lubineau, G. Ultrasensitive, stretchable strain sensors based on fragmented carbon nanotube papers. *ACS Appl. Mater. Interfaces* **2017**, *9* (5), 4835–4842.

[36]   Kang, D.; Pikhitsa, P. V.; Choi, Y. W.; Lee, C.; Shin, S. S.; Piao, L.; Park, B.; Suh, K. Y.; Kim, T. I.; Choi, M. Ultrasensitive mechanical crack-based sensor inspired by the spider sensory system. *Nature* **2014**, *516* (7530), 222–226.

[37]   Park, B.; Kim, J.; Kang, D.; Jeong, C.; Kim, K. S.; Kim, J. U.; Yoo, P. J.; Kim, T. I. Dramatically enhanced mechanosensitivity and signal-to-noise ratio of nanoscale crack-based sensors: effect of crack depth. *Adv. Mater.* **2016**, *28* (37), 8130–8137.

[38]   Park, B.; Lee, Y.; Jung, W.; Scott, D. K.; Aalto, D.; Chung, H. J.; Kim, T. I.; Deterministically assigned directional sensing of nanoscale crack based pressure sensor by anisotropic poisson ratios of the substrate. *J. Mater. Chem. C* **2021**, *9*, 5154–5161.

[39]   Liao, X.; Zhang, Z.; Kang, Z.; Gao, F.; Liao, Q.; Zhang, Y. Ultrasensitive and stretchable resistive strain sensors designed for wearable electronics. *Mater. Horiz.* **2017**, *4* (3), 502–510.

[40]   Sun, H.; Fang, X.; Fang, Z.; Zhao, L.; Tian, B.; Verma, P.; Maeda, R.; Jiang, Z. An ultrasensitive and stretchable strain sensor based on a microcrack structure for motion monitoring. *Microsyst. Nanoeng.* **2022**, *8*, 111.

[41]   Chen, S.; Wei, Y.; Wei, S.; Lin, Y.; Liu, L. Ultrasensitive cracking-assisted strain sensors based on silver nanowires/graphene hybrid particles. *ACS Appl. Mater. Interfaces* **2016**, *8* (38), 25563–25570.

[42] Wu, S.; Moody, K.; Kollipara, A.; Zhu, Y. Highly sensitive, stretchable, and robust strain sensor based on crack propagation and opening. *ACS Appl. Mater. Interfaces* **2023**, *15* (1), 1798–1807.

[43] Lipomi, D. J.; Vosgueritchian, M.; Tee, BCK.; Hellstrom, S. L.; Lee, J. A.; Fox, C. H.; Bao, Z. Skin-like pressure and strain sensors based on transparent elastic films of carbon nanotubes. *Nat. Nanotechnol.* **2011**, *6* (12), 788–792.

[44] Silva-López, M.; Fender, A.; Macpherson, W. N.; Barton, J. S.; Jones, JDC.; Zhao, D.; Dobb, H.; Webb, D. J.; Zhang, L.; Bennion, I. Strain and temperature sensitivity of a single-mode polymer optical fiber. *Opt. Lett.* **2005**, *30* (23), 3129–3131.

[45] Bai, H.; Li, S.; Barreiros, J.; Tu, Y.; Pollock, C. R.; Shepherd, R. F. Stretchable distributed fiber-optic sensors. *Science* **2020**, *370* (6518), 848–852.

[46] Tao, L. Q.; Wang, D. Y.; Tian, H.; Ju, Z. Y.; Liu, Y.; Pang, Y.; Chen, Y. Q.; Yang, Y.; Ren, T. L. Self-adapted and tunable graphene strain sensors for detecting both subtle and large human motions. *Nanoscale* **2017**, *9* (24), 8266–8273.

[47] Lee, J.; Shin, S.; Lee, S.; Song, J.; Kang, S.; Han, H.; Kim, S.; Kim, S.; Seo, J.; Kim, D.; Lee, T. Highly sensitive multifilament fiber strain sensors with ultrabroad sensing range for textile electronics. *ACS Nano* **2018**, *12* (5), 4259–4268.

[48] Lee, H.; Glasper, M. J.; Li, X.; Nychka, J. A.; Batcheller, J.; Chung, H. J.; Chen, Y. Preparation of fabric strain sensor based on graphene for human motion monitoring. *J. Mater. Sci.* **2018**, *53* (12), 9026–9033.

[49] Yin, F.; Ye, D.; Zhu, C.; Qiu, L.; Huang, Y. A. Stretchable, highly durable ternary nanocomposite strain sensor for structural health monitoring of flexible aircraft. *Sensors* **2017**, *17* (11), 2677.

[50] Aygun, L. E.; Kumar, V.; Weaver, C.; Gerber, M.; Wagner, S.; Verma, N.; Glisic, B.; Sturm, J. C. Large-area resistive strain sensing sheet for structural health monitoring. *Sensors* **2020**, *20* (5), 1386.

[51] Totland, C.; Thomas, P. J.; Størdal, I. F.; Eek, E. A. Fully distributed fibre optic sensor for the detection of liquid hydrocarbons. *IEEE Sens. J.* **2021**, *21* (6), 7631–7637.

[52] Cheng, X.; Bao, C.; Wang, X.; Dong, W. Stretchable strain sensor based on conductive polymer for structural health monitoring of high-speed train head. *Proc. Inst. Mech. Eng. L* **2020**, *234* (3), 496–503.

[53] Geetha, S.; Kumar, KKS.; Rao, CRK.; Trivedi, VDC. EMI shielding: Methods and materials—a review. *J. Appl. Polym. Sci.* **2009**, *112* (4); 2073–2086.

[54] Mathur, P.; Raman, S. Electromagnetic interference (EMI): measurement and reduction techniques. *J. Electron. Mater.* **2020**, *49*, 2975–2998.

[55] Lapinsky, S. E.; Easty, A. C. Electromagnetic interference in critical care. *J. Crit. Care* **2006**, *21* (3), 267–270.

[56] Jaroszewski, M.; Thomas, S.; Rane, A. V. *Advanced materials for electromagnetic shielding: fundamentals, properties, and applications*. Wiley, 2018.

[57] Iqbal, A.; Sambyal, P.; Koo, C. M. 2D MXenes for electromagnetic shielding: a review. *Adv. Funct. Mater.* **2020**, *30* (47), 2000883.

[58] Yun, G.; Tang, S. Y.; Lu, H.; Zhang, S.; Dickey, M. D.; Li, W. Hybrid-filler stretchable conductive composites: from fabrication to application. *Small Sci.* **2021**, *1* (6), 2000080.

[59] Tee, B. C.; Ouyang, J. Soft electronically functional polymeric composite materials for a flexible and stretchable digital future. *Adv. Mater.* **2018**, *30* (47), 1802560.

[60] Liu, C.; Cai, J.; Dang, P.; Li, X.; Zhang, D. Highly stretchable electromagnetic interference shielding materials made with conductive microcoils confined to a honeycomb structure. *ACS Appl. Mater. Interfaces* **2020**, *12* (10), 12101–12108.

[61] Song, S.; Xu, G.; Wang, B.; Liu, D.; Ren, Z.; Wang, C.; Zhao, J.; Zhang, L.; Li, Y. A multifunctional flexible electronic skin for dynamic thermal radiation regulation and electromagnetic interference shielding. *ACS Appl. Mater. Interfaces* **2022**, *14* (46), 52379–52389.

[62] Satish Kumar, K.; Rengaraj, R.; Venkatakrishnan, G. R.; Chandramohan, A. Polymeric materials for electromagnetic shielding—a review. *Mater. Today Proc.* **2021**, *47*, 4925–4928.

[63] Li, X.; Sun, X.; Zhang, J.; Xue, S.; Zhi, L. A stretchable fabric as strain sensor integrating electromagnetic shielding and electrochemical energy storage. *Nano Res.* **2023**, *16*, 12753–12761.

[64] Thomassin, J. M.; Jerome, C.; Pardoen, T.; Bailly, C.; Huynen, I.; Detrembleur, C. Polymer/carbon based composites as electromagnetic interference (EMI) shielding materials. *Mater. Sci. Eng. Rep.* **2013**, *74* (7), 211–232.

[65] Feng, P.; Ye, Z.; Wang, Q.; Chen, Z.; Wang, G.; Liu, X.; Li, K.; Zhao, W. Stretchable and conductive composites film with efficient electromagnetic interference shielding and absorptivity. *J. Mater. Sci.* **2020**, *55* (20); 8576–8590.

[66] Das, P.; Katheria, A.; Ghosh, S.; Roy, B.; Nayak, J.; Nath, K.; Paul, S.; Das, N. C. Self-healable and super-stretchable conductive elastomeric nanocomposites for efficient thermal management characteristics and electromagnetic interference shielding. *Synth. Met.* **2023**, *294*, 117304.

[67] Mani, D.; Vu, M. C.; Lim, C. S.; Kim, J. B.; Jeong, T. H.; Kim, H. J.; Islam, M. A.; Lim, J. H.; Kim, K. M.; Kim, S. R. Stretching induced alignment of graphene nanoplatelets in polyurethane films for superior in-plane thermal conductivity and electromagnetic interference shielding. *Carbon* **2023**, *201*, 568–576.

[68] Ma, R. Y.; Yi, S. Q.; Zhang, J. L.; Sun, W. J.; Jia, L. C.; Yan, D. X.; Li, Z. M. Highly efficient electromagnetic interference shielding and superior mechanical performance of carbon nanotube/polydimethylsiloxane composite with interface-reinforced segregated structure. *Compos. Sci. Technol.* **2023**, *232*, 109874.

[69] Hu, L.; Chee, P. L.; Sugiarto, S.; Yu, Y.; Shi, C.; Yan, R.; Yao, Z.; Shi, X.; Zhi, J.; Kai, D.; Yu, H. D.; Huang, W. Hydrogel-based flexible electronics. *Adv. Mater.* **2023**, *35* (14), 2205326.

[70] Zhu, Y.; Liu, J.; Guo, T.; Wang, J. J.; Tang, X.; Nicolosi, V. Multifunctional Ti_3C_2T x MXene composite hydrogels with strain sensitivity toward absorption-dominated electromagnetic-interference shielding. *ACS Nano* **2021**, *15* (1), 1465–1474.

[71] Zhao, B.; Bai, Z.; Lv, H.; Yan, Z.; Du, Y.; Guo, X.; Zhang, J.; Wu, L.; Deng, J.; Zhang, D. W.; Che, R. Self-healing liquid metal magnetic hydrogels for smart feedback sensors and high-performance electromagnetic shielding. *Nano-Micro Lett.* **2023**, *15* (1), 79.

[72] Jalali, A.; Zhang, R.; Rahmati, R.; Nofar, M.; Sain, M.; Park, C. B. Recent progress and perspective in additive manufacturing of EMI shielding functional polymer nanocomposites. *Nano Res.* **2023**, *16* (1); 1–17.

[73] Liu, J.; Mckeon, L.; Garcia, J.; Pinilla, S.; Barwich, S.; Mobius, M.; Stamenov, P.; Coleman, J. N.; Nicolosi, V. Additive manufacturing of Ti_3C_2-MXene-functionalized conductive polymer hydrogels for electromagnetic-interference shielding. *Adv. Mater.* **2022**, *34* (5), 2106253.

[74] Shi, S.; Peng, Z.; Jing, J.; Yang, L.; Chen, Y. 3D printing of delicately controllable cellular nanocomposites based on polylactic acid incorporating graphene/carbon nanotube hybrids for efficient electromagnetic interference shielding. *ACS Sustain. Chem. Eng.* **2020**, *8* (21), 7962–7972.

[75] Yang, Y.; Wang, J.; Song, C.; Pei, R.; Purushothama, J.; Zhang, Y. Electromagnetic shielding using flexible embroidery metamaterial absorbers: design, analysis and experiments. *Mater. Des.* **2022**, *222*, 111079.

[76] Almirall, O.; Fernández-García, R.; Gil, I.; Wearable metamaterial for electromagnetic radiation shielding. *J. Text. Inst.* **2022**, *113* (8), 1586–1594.

[77] Sakovsky, M.; Negele, J.; Costantine, J. Electromagnetic reconfiguration using stretchable mechanical metamaterials. *Adv. Sci.* **2023**, *10* (6), 2203376.

[78] Lee, N.; Lim, J. S.; Chang, I.; Bae, H. H.; Nam, J.; Cho, H. H. Flexible assembled metamaterials for infrared and microwave camouflage. *Adv. Opt. Mater.* **2022**, *10* (11), 2200448.

[79] Nan, Z.; Wei, W.; Lin, Z.; Chang, J.; Hao, Y. Flexible nanocomposite conductors for electromagnetic interference shielding. *Nano-Micro Lett.* **2023**, *15*, 172.

[80] Li, Y.; Tian, X.; Gao, S. P.; Li, K.; Yang, H.; Fu, F.; Lee, J. Y.; Guo, Y. X.; Ho, J. S.; Chen, P. Y. Reversible crumpling of 2D titanium carbide (MXene) nanocoatings for stretchable electromagnetic shielding and wearable wireless communication. *Adv. Funct. Mater.* **2020**, *30* (5), 1907451.

[81]   Xue, B.; Li, Y.; Cheng, Z.; Yang, S.; Xie, L.; Qin, S.; Zheng, Q. Directional electromagnetic interference shielding based on step-wise asymmetric conductive networks. *Nano-Micro Lett.* **2022**, *14*, 1–16.

[82]   Pu, J. H.; Zha, X. J.; Tang, L. S.; Bai, L.; Bao, R. Y.; Liu, Z. Y.; Yang, M. B.; Yang, W. Human skin-inspired electronic sensor skin with electromagnetic interference shielding for the sensation and protection of wearable electronics. *ACS Appl. Mater. Interfaces* **2018**, *10* (47), 40880–40889.

[83]   Liu, J.; Lin, S.; Huang, K.; Jia, C.; Wang, Q.; Li, Z.; Song, J.; Liu, Z.; Wang, H.; Lei, M.; Wu, H. A large-area AgNW-modified textile with high-performance electromagnetic interference shielding. *npj Flex. Electron.* **2020**, *4* (1), 10.

[84]   Fan, S. T.; Guo, D. L.; Zhang, Y. T.; Chen, T.; Li, B. J.; Zhang, S. Washable and stable coaxial electrospinning fabric with superior electromagnetic interference shielding performance for multifunctional electronics. *Chem. Eng. J.* **2024**, *488*, 151051.

[85]   Zhou, Y.; Dong, X.; Mi, Y.; Fan, F.; Xu, Q.; Zhao, H.; Wang, S.; Long, Y. Hydrogel smart windows. *J. Mater. Chem. A* **2020**, *8* (20), 10007–10025.

[86]   Pérez-Lombard, L.; Ortiz, J.; Pout, C. A review on buildings energy consumption information. *Energy Build.* **2008**, *40* (3), 394–398.

[87]   Khandelwal, H.; Schenning, A. P.; Debije, M. G. Infrared regulating smart window based on organic materials. *Adv. Energy Mater.* **2017**, *7* (14), 1602209.

[88]   Smith, G.; Gentle, A.; Arnold, M.; Cortie, M. Nanophotonics-enabled smart windows, buildings and wearables. *Nanophotonics* **2016**, *5* (1), 55–73.

[89]   Kammen, D. M.; Sunter, D. A. City-integrated renewable energy for urban sustainability. *Science* **2016**, *352* (6288), 922–928.

[90]   Shchegolkov, A. V.; Jang, S. H.; Shchegolkov, A. V.; Rodionov, Y. V.; Sukhova, A. O.; Lipkin, M. S. A brief overview of electrochromic materials and related devices: A nanostructured materials perspective. *Nanomaterials* **2021**, *11* (9), 2376.

[91]   Svensson, J. S. E. M.; Granqvist, C. G. Electrochromic coatings for "smart windows". *Sol. Energy Mater.* **1985**, *12* (6), 391–402.

[92]   Svensson, J. S. E. M.; Granqvist, C. G. Electrochromic coatings for smart windows: Crystalline and amorphous WO_3 films. *Thin Solid Films* **1985**, *126*, 31–36.

[93]   Sala, R. L.; Goncalves, R. H.; Camargo, E. R.; Leite, E. R. Thermosensitive poly (N-vinylcaprolactam) as a transmission light regulator in smart windows. *Sol. Energy Mater. Sol. Cells* **2018**, *186*, 266–272.

[94]   Wang, S.; Owusu, K. A.; Mai, L.; Ke, Y.; Zhou, Y.; Hu, P.; Magdassi, S.; Long, Y. Vanadium dioxide for energy conservation and energy storage applications: synthesis and performance improvement. *Appl. Energy* **2018**, *211*, 200–217.

[95]   Azens, A.; Granqvist, C. Electrochromic smart windows: energy efficiency and device aspects. *J. Solid State Electrochem.* **2003**, *7*, 64–68.

[96]   Wang, J. L.; Sheng, S. Z.; He, Z.; Wang, R.; Pan, Z.; Zhao, H. Y.; Liu, J. W.; Yu, S. H. Self-powered flexible electrochromic smart window. *Nano Lett.* **2021**, *21* (23), 9976–9982.

[97]   Pardo, R.; Zayat, M.; Levy, D. Photochromic organic–inorganic hybrid materials. *Chem. Soc. Rev.* **2011**, *40* (2), 672–687.

[98]   Nandakumar, D. K.; Ravi, S. K.; Zhang, Y.; Guo, N.; Zhang, C.; Tan, S. C. A super hygroscopic hydrogel for harnessing ambient humidity for energy conservation and harvesting. *Energy Environ. Sci.* **2018**, *11* (8), 2179–2187.

[99]   Zhou, Y.; Fan, F.; Liu, Y.; Zhao, S.; Xu, Q.; Wang, S.; Luo, D.; Long, Y. Unconventional smart windows: materials, structures and designs. *Nano Energy* **2021**, *90*, 106613.

[100] Wang, X.; Narayan, S. Thermochromic materials for smart windows: a state-of-art review. *Front. Energy Res.* **2021**, *9*, 800382.

[101] La, T. G.; Li, X.; Kumar, A.; Fu, Y.; Yang, S.; Chung, H. J. Highly flexible, multipixelated thermosensitive smart windows made of tough hydrogels. *ACS Appl. Mater. Interfaces* **2017**, *9* (38), 33100–33106.

[102] Saeli, M.; Piccirillo, C.; Parkin, I. P.; Binions, R.; Ridley, I. Energy modelling studies of thermochromic glazing. *Energy Build.* **2010**, *42* (10), 1666–1673.

[103] Cui, Y.; Ke, Y.; Liu, C.; Chen, Z.; Wang, N.; Zhang, L.; Zhou, Y.; Wang, S.; Gao, Y.; Long, Y. Thermochromic VO_2 for energy-efficient smart windows. *Joule* **2018**, *2* (9), 1707–1746.

[104] Xu, Y.; Huang, W.; Shi, Q.; Zhang, Y.; Song, L.; Zhang, Y. Synthesis and properties of Mo and W ions co-doped porous nano-structured VO_2 films by sol–gel process. *J. Sol-Gel Sci. Technol.* **2012**, *64*, 493–499.

[105] Zhou, J.; Gao, Y.; Zhang, Z.; Luo, H.; Cao, C.; Chen, Z.; Dai, L.; Liu, X. VO_2 thermochromic smart window for energy savings and generation. *Sci. Rep.* **2013**, *3* (1), 3029.

[106] Zhang, S.; Cao, S.; Zhang, T.; Lee, J. Y. Plasmonic oxygen-deficient TiO_{2-x} nanocrystals for dual-band electrochromic smart windows with efficient energy recycling. *Adv. Mater.* **2020**, *32* (43), 2004686.

[107] Cai, G.; Cui, M.; Kumar, V.; Darmawan, P.; Wang, J.; Wang, X.; Eh, A. L. S.; Qian, K.; Lee, P. S. Ultra-large optical modulation of electrochromic porous WO_3 film and the local monitoring of redox activity. *Chem. Sci.* **2016**, *7* (2), 1373–1382.

[108] Wu, L. Y.; Zhao, Q.; Huang, H.; Lim, R. J. Sol-gel based photochromic coating for solar responsive smart window. *Surf. Coat. Technol.* **2017**, *320*, 601–607.

[109] Wu, M.; Shi, Y.; Li, R.; Wang, P. Spectrally selective smart window with high near-infrared light shielding and controllable visible light transmittance. *ACS Appl. Mater. Interfaces* **2018**, *10* (46), 39819–39827.

[110] Yamazaki, S.; Ishida, H.; Shimizu, D.; Adachi, K. Photochromic properties of tungsten oxide/methylcellulose composite film containing dispersing agents. *ACS Appl. Mater. Interfaces* **2015**, *7* (47), 26326–26332.

[111] Xing, Z.; Jia, X.; Li, X.; Yang, J.; Wang, S.; Li, Y.; Shao, D.; Feng, L.; Song, H. Novel green reversible humidity-responsive hemiaminal dynamic covalent network for smart window. *ACS Appl. Mater. Interfaces* **2023**, *15* (8), 11053–11061.

[112] Cai, G.; Wang, J.; Eh, ALS.; Chen, J.; Qian, K.; Xiong, J.; Thangavel, G.; Lee, P. S. Diphylleia grayi-inspired stretchable hydrochromics with large optical modulation in the visible–near-infrared region. *ACS Appl. Mater. Interfaces* **2018**, *10* (43), 37685–37693.

[113] Cotur, Y.; Olenik, S.; Asfour, T.; Bruyns-Haylett, M.; Kasimatis, M.; Tanriverdi, U.; Gonzalez-Macia, L.; Lee, H. S.; Kozlov, A. S.; Güder, F. Bioinspired stretchable transducer for wearable continuous monitoring of respiratory patterns in humans and animals. *Adv. Mater.* **2022**, *34*, 2203310.

[114] Chang, T.; Akin, S.; Kim, M. K.; Murray, L.; Kim, B.; Cho, S.; Huh, S.; Teke, S.; Couetil, J.; Jun, M. B. G.; Lee, C. H. A programmable dual-regime spray for large-scale and custom-designed electronic textiles. *Adv. Mater.* **2022**, *34*, 2108021.

[115] Nassar, J. M.; Khan, S. M.; Velling, S. J.; Diaz-Gaxiola, A.; Shaikh, S. F.; Geraldi, N. R.; Torres Sevilla, G. A.; Duarte, C. M.; Hussain, M. M. Compliant lightweight non-invasive standalone "Marine Skin" tagging system. *npj Flex. Electron.* **2018** *2*, 13.

[116] Zhang, L.; Li, Y.; Du, J.; Mu, B.; Hu, J.; Zhang, X. Flexible bioimpedance-based dynamic monitoring of stress levels in live oysters. *Aquaculture* **2023**, *577*, 739957.

[117] Oren, S.; Ceylan, H.; Schnable, P. S.; Dong, L. High-resolution patterning and transferring of graphene-based nanomaterials onto tape toward roll-to-roll production of tape-based wearable sensors. *Adv. Mater. Technol.* **2017**, *2*, 1700223.

[118] Zhang, C.; Zhang, C.; Wu, X.; Ping, J.; Ying, Y. An integrated and robust plant pulse monitoring system based on biomimetic wearable sensor. *npj Flex. Electron.* **2022**, *6*, 43.

[119] Li, Z.; Liu, Y.; Hossain, O.; Paul, R.; Yao, S.; Wu, S.; Ristaino, J.; Zhu, Y.; Wei, Q. Real-time monitoring of plant stresses via chemiresistive profiling of leaf volatiles by a wearable sensor. *Matter* **2021**, *4*, 2553–2570.

[120] Lee, G.; Hossain, O.; Jamalzadegan, S.; Liu, Y.; Wang, H.; Saville, A. C.; Shymanovich, T.; Paul, R.; Rotenberg, D.; Whitfield, A. E.; Ristaino, J. B.; Zhu, Y.; Wei, Q. Abaxial leaf surface-mounted multimodal wearable sensor for continuous plant physiology monitoring. *Sci. Adv.* **2023**, *9*, eade2232.

[121] Basodan, R. A.; Park, B.; Chung, H. J. Smart personal protective equipment (PPE): current PPE needs, opportunities for nanotechnology and e-textiles. *Flex. Print. Electron.* **2022**, *6* (4), 043004.

[122] Zhang, L. S.; Li, J.; Wang, F.; Shi, J. D.; Chen, W.; Tao, X. M. Flexible stimuli-responsive materials for smart personal protective equipment. *Mater. Sci. Eng. Rep.* **2021**, *146*, 100629.

[123] Saidi, A.; Gauvin, C.; Ladhari, S.; Nguyen-Tri, P. Advanced functional materials for intelligent thermoregulation in personal protective equipment. *Polymers* **2021**, *13* (21), 3711.

[124] Kim, J. J.; Wang, Y.; Wang, H.; Lee, S.; Yokota, T.; YSomeya, T. Skin electronics: next-generation device platform for virtual and augmented reality. *Adv. Funct. Mater.* **2021**, *31* (39), 2009602.

[125] Biswas, S.; Visell, Y. Haptic perception, mechanics, and material technologies for virtual reality. *Adv. Funct. Mater.* **2021**, *31* (39), 2008186.

[126] Kim, H.; Kwon, Y. T.; Lim, H. R.; Kim, J. H.; Kim, Y. S.; Yeo, W. H. Recent advances in wearable sensors and integrated functional devices for virtual and augmented reality applications. *Adv. Funct. Mater.* **2021**, *31* (39), 2005692.

[127] Ko, S. H.; Rogers, J. Functional materials and devices for XR (VR/AR/MR) applications. *Adv. Funct. Mater.* **2021**, *31* (39), 2106546.

[128] Brown, C.; Hicks, J.; Rinaudo, C. H.; Burch, R. The use of augmented reality and virtual reality in ergonomic applications for education, aviation, and maintenance. *Ergon. Des.* **2023**, *31* (4), 23–31.

[129] Chen, H.; Hou, L.; Zhang, G. K.; Moon, S. Development of BIM, IoT and AR/VR technologies for fire safety and upskilling. *Autom. Constr.* **2021**, *125*, 103631.

[130] Braun, P.; Grafelmann, M.; Gill, F.; Stolz, H.; Hinckeldeyn, J.; Lange, A. K. Virtual reality for immersive multi-user firefighter training scenarios. *Virtual Real Intell. Hardw.* **2022**, *4* (5), 406–417.

[131] Ma, L.; Mor, S.; Anderson, P. L.; Baños, R. M.; Botella, C.; Bouchard, S.; Cárdenas-López, G.; Donker, T.; Frenandez-Alverez, J.; Lindner, P.; Muhlberger, A.; Powers, M. B.; Quero, S.; Rothbaum, B.; Wiederhold, B. K.; Carbring, P. Integrating virtual realities and psychotherapy: SWOT analysis on VR and MR based treatments of anxiety and stress-related disorders. *Cogn. Behav. Ther.* **2021**, *50* (6), 509–526.

[132] Hinze, J.; Röder, A.; Menzie, N.; Müller, U.; Domschke, K.; Riemenschneider, M.; Noll-Hussong, M. Spider phobia: neural networks informing diagnosis and (virtual/augmented reality-based) cognitive behavioral psychotherapy—a narrative review. *Front. Psychiatry* **2021**, *12*, 704174.

[133] Kim, H.; Kwon, Y. T.; Lim, H. R.; Kim, J. H.; Kim, Y. S.; Yeo, W. H. Recent advances in wearable sensors and integrated functional devices for virtual and augmented reality applications. *Adv. Funct. Mater.* **2021**, *31* (39), 2005692.

[134] Bai, H.; Li, S.; Shepherd, R. F. Elastomeric haptic devices for virtual and augmented reality. *Adv. Funct. Mater.* **2021**, *31* (39), 20.

[135] Jung, Y. H.; Kim, J. H.; Rogers, J. A. Skin-integrated vibrohaptic interfaces for virtual and augmented reality. *Adv. Funct. Mater.* **2021**, *31* (39), 2008805. 09364.

[136] Forsyth, J. B.; Martin, T. L.; Young-Corbett, D.; Dorsa, E. Feasibility of intelligent monitoring of construction workers for carbon monoxide poisoning. *IEEE Trans. Autom. Sci. Eng.* **2012**, *9* (3), 505–515.

[137] Kozlovszky, M.; Pavlinić, D. Z. Environment and situation monitoring for firefighter teams. In *IEEE 15th international symposium on computational intelligence and informatics (CINTI)*; 2014; pp. 439–442.

[138] Calabrò, R. S.; Cerasa, A.; Ciancarelli, I.; Pignolo, L.; Tonin, P.; Iosa, M.; Morone, G. The arrival of the metaverse in neurorehabilitation: fact, fake or vision?. *Biomedicines* **2022**, *10* (10), 2602.

[139] Yang, X.; Yu, Y.; Shirowzhan, S.; Li, H. Automated PPE-tool pair check system for construction safety using smart IoT. *J. Build. Eng.* **2020**, *32*, 101721.

[140] Chang, J. S. K.; Henry, M. J.; Burtner, R.; Love, O.; Corley, C. The heroes' problems: exploring the potentials of Google Glass for biohazard handling professionals. In *Proceedings of the 33rd annual acm conference extended abstracts on human factors in computing systems*; 2015; pp. 1531–1536.

[141] Buller, M. J.; Tharion, W. J.; Duhamel, C. M.; Yokota, M. Real-time core body temperature estimation from heart rate for first responders wearing different levels of personal protective equipment. *Ergonomics* **2015**, *58* (11), 1830–1841.

[142] Zhou, Y.; He, J.; Wang, H.; Qi, K.; Nan, N.; You, X.; Shao, W.; Wang, L.; Ding, B.; Cui, S. Highly sensitive, self-powered and wearable electronic skin based on pressure-sensitive nanofiber woven fabric sensor. *Sci. Rep.* **2017**, *7* (1), 12949.

[143] Rasouli, S.; Alipouri, Y.; Chamanzad, S. Smart personal protective equipment (PPE) for construction safety: a literature review. *Saf. Sci.* **2024**, *170*, 106368.

[144] De Fazio, R.; Mastronardi, V. M.; Petruzzi, M.; De Vittorio, M.; Visconti, P. Human–machine interaction through advanced haptic sensors: a piezoelectric sensory glove with edge machine learning for gesture and object recognition. *Future Internet* **2022**, *15* (1), 14.

[145] Lee, J.; Sul, H.; Lee, W.; Pyun, K. R.; Ha, I.; Kim, D.; Park, H.; Eom, H.; Yoon, Y.; Jung, J.; Lee, D.; Ko, S. H. Stretchable skin-like cooling/heating device for reconstruction of artificial thermal sensation in virtual reality. *Adv. Funct. Mater.* **2020**, *30* (29), 1909171.

[146] Mahmood, M.; Kim, N.; Mahmood, M.; Kim, H.; Kim, H.; Rodeheaver, N.; Sang, M.; Yu, K. J.; Yeo, W. H. VR-enabled portable brain-computer interfaces via wireless soft bioelectronics. *Biosens. Bioelecton.* **2022**, *210*, 114333.

[147] Sun, Z.; Zhu, M.; Shan, X.; Lee, C. Augmented tactile-perception and haptic-feedback rings as human–machine interfaces aiming for immersive interactions. *Nat. Commun.* **2022**, *13* (1), 5224.

[148] Xu, K.; Lu, Y.; Takei, K. Flexible hybrid sensor systems with feedback functions. *Adv. Funct. Mater.* **2021**, *31* (39), 2007436.

[149] Fang, H.; Guo, J.; Wu, H. Wearable triboelectric devices for haptic perception and VR/AR applications. *Nano Energy* **2022**, *96*, 107112.

[150] Lee, J.; Kim, D.; Sul, H.; Ko, S. H. Thermo-haptic materials and devices for wearable virtual and augmented reality. *Adv. Funct. Mater.* **2021**, *31* (39), 2007376.

[151] Oh, J.; Kim, S.; Lee, S.; Jeong, S.; Ko, S. H.; Bae, J. A liquid metal based multimodal sensor and haptic feedback device for thermal and tactile sensation generation in virtual reality. *Adv. Funct. Mater.* **2021**, *31* (39), 2007772.

[152] Zou, M.; Li, S.; Hu, X.; Leng, X.; Wang, R.; Zhou, X.; Liu, Z. Progresses in tensile, torsional, and multifunctional soft actuators. *Adv. Funct. Mater.* **2021**, *31* (39), 2007437.

Tricia Breen Carmichael and Hyun-Joong Chung

Stretchable electronics: the next generation of emerging applications

The field of stretchable electronics has evolved at a tremendous pace over the past 20 years. Chapter 1 described the origins of the field, with the first *structures that stretch*, the serpentine interconnects with an island–bridge architecture, published in 2004. The interdisciplinary pursuit of 'the structures that stretch' rapidly developed designs that became increasingly sophisticated, with form-factor innovations expanding to thin films, meshes, fibers, tissue hybrids, and textiles. These innovations also incorporated techniques used in the creative arts, such as origami and kirigami, to manufacture 3D structures. Chapter 1 also covered the parallel track—the pursuit of *materials that stretch*—that transform brittle conductors and semiconductors into rubber-like forms via compositing or copolymerization. The first stretchable functional material with metal-level electrical conductivity, EGaIn liquid metal in elastomeric packaging as a moldable soft electrode, was published in 2008. Material innovations in this area quickly expanded to include hydrogels, conjugated polymers, supramolecular polymers, nanocomposites, and bioresorbable inorganics.

After a mere two decades, stretchable interconnects merged with miniaturized sensors, actuators, and data processors to make meaningful progress, both in academia and industrial sectors, in personalized healthcare, human–machine interfaces, and soft robotics. Stretchable conductor and interconnect technologies have been of prime importance because stretchable yet low-impedance wiring is the most fundamental building block that can enable multiplexed arrays of sensors over a wide area of irregularly shaped surfaces. The immense amount of collected data are stored and processed by cloud computing. Efficient data-collection processes can be enhanced by neuromorphic computing. A good example that combines these technologies may be electronic skins for prosthesis, where electronics with rubber-like form factors can integrate with irregularly shaped moving substances in a seamless manner [1]. Stretchable electronics lies at the center of the future of electronics, and it spearheading the value-added electronics revolution called "More than Moore" (MtM).

In documenting the immensely wide and diverse field of stretchable electronics, Chapters 2 to 5 were designed to collect the visions and voices of a young generation of principal investigators in the world, specifically with independent research experience of 10 years or a similar extent. These investigators have brought unconventional ideas to the field of stretchable electronics and have contributed to the enrichment of

Tricia Breen Carmichael, Department of Chemistry and Biochemistry, University of Windsor, 401 Sunset Ave., Windsor, Ontario N9B 3P4, Canada, e-mail: tbcarmic@uwindsor.ca
Hyun-Joong Chung, Department of Chemical and Materials Engineering, University of Alberta, Edmonton, Alberta T6G 1H9, Canada, e-mail: chung3@ualberta.ca

https://doi.org/10.1515/9783110757286-006

this vibrant field. As a result, unforeseen engineering applications have been discovered by brilliant young investigators the world over. The field has evolved to become much more diverse than electronic devices simply being stretchable. At the same time, the interdisciplinary field of stretchable electronics may seem daunting to aspiring junior researchers who are about to pursue their own independent careers; it is hoped that the chapter contributors' down-to-earth approaches to designing their respective research programs—while surviving the rigors of early career research—can be inspirational.

Chapters 2 and 3 were stories about *functional materials that stretch*. In Chapter 2, Nyayachavadi and Rondeau–Gagné introduced cross-linking strategies to solve the most challenging problem of the field of organic π-conjugated polymer-based technologies, namely, long-term stability. The field of cross-linking chemistry is simultaneously fundamental and applied. Like how vulcanization of natural rubber by sulfur revolutionized automobile industry, the cross-linking of π-conjugated polymers can be a game changer in the electronics industry. In Chapter3, Barron and Bartlett introduced the fabrication of liquid metal-based electronic composites and their application in soft robotics. Rapid oxide formation and the high surface energy of liquid metals has enabled soft robotic devices with remarkable form-factor adaptability with robustness. The future of liquid metal-based composites looks even more promising with the incorporation of magnetic particles. The efficacy of these novel stretchable functional materials looks extremely promising based on the soft robotics devices that the authors showcased.

Chapter 4 was a unique application story of *structures that stretch*; perhaps, this application is the most promising among all the stretchable electronics derived technologies. Fish and crustacean eyes have fascinating ocular structures with detector arrays (rods and cones in the retina) on the inner surface of the eyeball. This form factor is far more effective than the planar detector arrays of a camera in terms of obtaining distortion-free images without aberration. In the chapter, Chang, Yeo, and Song showcased stretchable photodetectors and image sensors that mimic the structure and tunability of fish and crustacean eyes.

Chapter 5 reminded the reader that the new technologies developed while pursuing futuristic applications of stretchable electronics can be applied to some of the most traditional engineering applications. In the chapter, Chung and coworkers narrated a case story about strain gauges, especially on how stretchable electronics-derived materials and structures rejuvenated the field that is more than a century old. The relationships between the stretchable technologies and other relevant fields of engineering, such as EMI shielding, smart windows, animal-care, and agriculture industry, were also discussed. The fields of traditional engineering have established supply chains with budgets with astronomical numbers, so even a miniscule influence can impact and benefit the world economy in a significant way.

The field of stretchable electronics has matured during the past 20 years, along the way training generations of researchers. This book presents the work and perspectives of this next generation of innovators and disruptors who are driving this exciting phase of diversification and growth in the field. The content coverage of this book is far from complete. Most notably, the evolution of applications in energy harvesting, storage, or wireless power-transmission devices are not covered. Nevertheless, we believe that the fresh and unique views of the young researchers showcased in this book can inspire and help guide avid readers who are looking for different perspectives in the crowded field of stretchable electronics.

Bibliography

[1] Luo, Y. et al. Technology roadmap for flexible sensors. *ACS Nano* **2023**, *17* (6), 5211–5295.

Abbreviations

Ag	Silver
EGaIn	Eutectic Gallium–Indium
FC	Magnetorheological fluid composite
Ga	Gallium
HC	Hybrid composite
In	Indium
LED	Light emitting diode
LM	Liquid metal
LMPA	Low melting-point alloy
MR	Magnetorheological
MRE	Magnetorheological elastomer
MRF	Magnetorheological fluid
RC	Rigid particle composite
SLICE	Shearing liquids into complex particles
SMA	Shape memory alloy
Sn	Tin
PD	Photodetector
NW	Nanowire
NM	Nanomembrane
PET	polyethylene terephthalate
PDMS	polydimethylsiloxane
PI	polyimide
0D	0-dimensional
1D	1-dimensional
2D	2-dimensional
QD	quantum dot
NP	Nanoparticle
NC	Nanocrystal
ITO	Indium tin oxide
TBAI	Tetrabutylammonium iodide
EDT	2-ethanedithiol
I-V	Current-voltage
TEM	Transmission electron microscopic
ETL	Electron transport layer
DFPBr-6	Fluorene pyridinium bromide derivative
J	Current density
V	Voltage
NT	Nanotube
VLS	Vapor–liquid–solid
VSS	Vapor–solid–solid
PP	Polypropylene
PVP	Polyvinylpyrrolidone
CVD	Chemical vapor deposition
PVP	Poly(vinylpyrrolidone)
PAM	Porous alumina membrane
VSSR	Vapor–solid–solid-reaction
TMD	Transition metal dichalcogenide

https://doi.org/10.1515/9783110757286-007

CMOS	Complementary metal-oxide semiconductor
GOI	Ge-on-insulator
PLD	Pulsed-laser deposition
HRTEM	High-resolution TEM
NR	Nanorod
CNT	Carbon nanotube
CAS	Conformal additive stamp
BHJ	Bulk heterojunction
TFT	Thin film transistor
PFM	Perovskite-filled membrane
ppi	Pixels per inch
LTPS	Low-temperature polycrystalline silicon
ECL	Edge cover layer
PPG	Photoplethysmogram
SO_2	Oxygen saturation
GQD	Graphene quantum dot
c-Si-PDA	Cylindrical silicon PD array
ROI	Region of interest
IC	integrated-circuit
EFPI	Extrinsic Fabry-Perot Interferometry
FBG	Fiber Bragg-Grating spectrometry
PMMA	Polymethyl methacrylate
POF	Plastic optical fiber
CNT	Carbon nanotube
CB	Carbon black
MWCNT	Multiwall carbon nanotubes
EM	Electromagnetic
EMI	Electromagnetic interference
SE	Shielding effectiveness
PDMS	Polydimethylsiloxane
GNP	Graphite nanoplates
PU	Polyurethane
MXene	Metallocene
HVAC	Heating, ventilation, and air conditioning
VO_2	Vanadium dioxide
ASiT	Air-silicone composite transducer
VOC	Volatile organic compound
rGO	Reduced graphene oxide
PPE	Personal protective equipment
AR	Augmented reality
VR	Virtual reality

Mathematical symbols

ΔE_c	Change in composite elastic modulus
ϵ_r	Relative permittivity
μ_{eff}	Effective relative permeability
μ_i	Inclusion relative permeability
μ_e	Elastomer relative permeability
σ	Electrical conductivity
ϕ	Inclusion volume fraction
Ψ	Magnetic fluid volume fraction
$\Omega_{\text{composite}}$	Magnetic particle volume fraction
k	Thermal conductivity
E_c	Composite elastic modulus
$E_{c,B=0}$	Zero-field composite elastic modulus
E_e	Elastomer elastic modulus
T_m	Melting point
T_b	Boiling point

https://doi.org/10.1515/9783110757286-008

Index

https://doi.org/10.1515/9783110757286-009